The Nuclear Spies

The Nuclear Spies

America's Atomic Intelligence Operation against Hitler and Stalin

Vince Houghton

Cornell University Press
Ithaca and London

First published 2019 by Cornell University Press

Printed in the United States of America

Library of Congress Cataloging-in-Publication Data

Names: Houghton, Vince, author.
Title: The nuclear spies : America's atomic intelligence operation against Hitler and Stalin / Vince Houghton.
Description: Ithaca : Cornell University Press, 2019. | Includes bibliographical references and index.
Identifiers: LCCN 2018052485 (print) | LCCN 2018052015 (ebook) | ISBN 9781501739590 (cloth : alk. paper) | ISBN 9781501739606 (pdf) | ISBN 9781501739613 (ret)
Subjects: LCSH: Espionage, American—Germany—History—20th century. | Espionage, American—Soviet Union—History. | Nuclear weapons information—History—20th century. | World War, 1939–1945—Military intelligence. | Cold War—Military intelligence. | Military intelligence—United States—History—20th century.
Classification: LCC UB271.U5 H68 2019 (ebook) | LCC UB271.U5 (print) | DDC 327.127304709/044—dc23
LC record available at https://lccn.loc.gov/2018052485

For Jon and Rae

The United States has come to see that it is in a new kind of rivalry with the Soviet Union—a rivalry that may well turn, not on territorial or diplomatic gains, or even (in the narrow sense of the word) on military advantage. The crucial advantage in the issue of power is likely to be with the nation whose scientific program can produce the next revolutionary advance in military tactics, following those already made by radar, jet propulsion, and nuclear fission.

—Don K. Price, *Government and Science*, 1954

For the whole world was flaring then into a monstrous phase of destruction. Power after Power about the armed globe sought to anticipate attack by aggression. They went to war in a delirium of panic, in order to use their bombs first. China and Japan had assailed Russia and destroyed Moscow, the United States had attacked Japan, India was in anarchistic revolt with Delhi a pit of fire spouting death and flame; the redoubtable King of the Balkans was mobilising. It must have seemed plain at last to every one in those days that the world was slipping headlong to anarchy. By the spring of 1959 from nearly two hundred centres, and every week added to their number, roared the unquenchable crimson conflagrations of the atomic bombs, the flimsy fabric of the world's credit had vanished, industry was completely disorganised and every city, every thickly populated area was starving or trembled on the verge of starvation. Most of the capital cities of the world were burning; millions of people had already perished, and over great areas government was at an end.

—H. G. Wells, *The World Set Free*, 1914

Contents

Introduction

The Principal Uncertainty

The storyline is well known, but not necessarily well understood. In September 1949, the U.S. intelligence establishment was shocked to discover that the Soviet Union had detonated its first atomic bomb. Coming just four years after the United States had become the world's first nuclear power, the Soviet atomic bomb was produced in half the time that U.S. intelligence had predicted. The consensus among the intelligence community, American scientists, the military, and the civilian political leadership had been that the earliest probable date for a Soviet atomic bomb was 1953. Somehow the Soviet Union had exceeded the expectations of U.S. national security experts by almost four years.

Compounding the confusion of U.S. leadership was the fact that, during the Second World War, U.S. intelligence had engaged in an effort against Nazi Germany that had correctly assessed the status of the German atomic bomb program. The German program had been given considerable attention by U.S. intelligence, yet despite the initial belief that the German atomic bomb project was significantly ahead of the progress of the United States' Manhattan Project, in 1944 U.S. intelligence discovered that the Germans would not develop an atomic bomb in time to affect the outcome of the war.

Both of these efforts operated within the framework of an entirely new field of intelligence: scientific intelligence. For the first time in history a nation's scientific resources—the abilities of its scientists, the state of its research institutions and laboratories, its scientific educational system—became a key consideration in assessing a potential national security threat. Information concerning a nation's technological capabilities had been a priority for U.S. intelligence organizations since the American Revolution. Yet scientific intelligence was a product of the Second World War and the development—and strategic implications—of the atomic bomb. The atomic bomb itself was a direct application of scientific theory to a weapon of war, the culmination of four decades of scientific research into the physics of the atom. Nuclear weapons, therefore, made intelligence about an enemy nation's scientific abilities an integral part of strategic planning. It was no longer sufficient to know just the ramifications of an enemy's deployed weapons systems or technological achievements. With the advent of a weapon of unprecedented destructive force, it became paramount to acquire information about an enemy's scientists, research laboratories, universities, and overall scientific infrastructure in order to correctly assess the immenseness of dire strategic threat. Such information was indeed crucial to national survival.

Scientific intelligence also forced a change in thinking about intelligence collection and analysis. Other types of intelligence can base their collection and analysis efforts on tangible things: technological intelligence can look at an aircraft and calculate the air speed, payload, survivability; military intelligence can count tanks, troops, divisions; economic intelligence can determine industrial capability, monitor debt, calculate GDP. Yet scientific intelligence is primarily focused on future potential, on how the scientific abilities of a particular state might, at some point, threaten national security. In doing so, scientific intelligence makes general assessments of a nation's scientific ability and presupposes that these findings are indicators of a potential strategic threat. In other words, assessments made about particular scientific research with strategic applications—such as the ability to develop nuclear weapons—are extrapolated from general assumptions made about the totality of a nation's scientific abilities.

In 1942 American scientists began to fear the possibility that Germany would develop and deploy an atomic bomb before the Manhattan Project could build its own weapon.[1] As a result, American scientists created an ad hoc organization for atomic intelligence, drawing on their scientific contacts in Europe and their scientific experience to learn and discern what they could about the German atomic bomb program. In the summer of 1943 the U.S. government authorized the Manhattan Project's director, Brig. Gen. Leslie Groves, to take complete control of all atomic intelligence–related operations. This action was a response to the acute fear of German scientific ability, the ineffectiveness of American scientists in collecting any actionable intelligence on their own, and the recognition that there were no existing intelligence organizations that could carry out such a difficult task. In doing so, the government gave Groves unprecedented power to centralize and consolidate intelligence functions.

A little over a year later, the decision to give Groves that responsibility paid off. The Manhattan Engineer District (MED—the formal name of the Manhattan Project) intelligence team discovered evidence that strongly indicated that the German atomic bomb program was significantly behind that of the United States, and thus it was highly unlikely that Germany would have an atomic bomb before the end of the war. The MED intelligence effort against the Germans was successful because it excelled at all three aspects of what is known as the "intelligence cycle": collection, analysis, and dissemination. Collection is the acquisition of information from a variety of sources such as human intelligence, signals intelligence, and imagery intelligence. Analysis is taking the raw data acquired by the collection effort and discerning its military significance. This is done through the creation of intelligence estimates, which assess the capabilities and intentions of a prospective opponent. Finally, dissemination is presenting this analysis to policymakers—and convincing them of its validity so that they can make use of it in the formation of national policies and strategies. A failure in any one aspect of the intelligence cycle means failure of the whole.

Leslie Groves's intelligence organization collected a wealth of information from a variety of sources. His analysis contingent was highly capable and quickly transformed the collected information into a conclusive argument. As a result, although the U.S. scientific, military, and political leadership had a deep-seated belief in the abilities of the German scientists, Groves's intelligence system was effective enough to convince the U.S. leadership that the German atomic program lagged behind that of the United States and that there would, as a consequence, be no German bomb.

When the Second World War ended, the United States had a capable atomic intelligence organization that had achieved great success against the Germans. Yet in the early postwar period, its institutional foundations were significantly weakened along with much of the rest of the U.S. intelligence apparatus. Despite the knowledge gained through the German experience, the available trained intelligence personnel, and the existing atomic intelligence organization, atomic intelligence against the Soviet nuclear program was not an immediate priority. Although the U.S. atomic intelligence apparatus would later be rebuilt, the rebuilding process was not done with a sense of urgency. Instead, American scientific and intelligence leaders assumed they had ample time—years, perhaps even decades—to create an effective system before the Soviets could build a bomb.

The result was an atomic intelligence system that failed in all three aspects of the intelligence cycle. Collection was done piecemeal, through a variety of intelligence organizations, and could not provide analysts with anything close to a complete picture of the status of the Soviet atomic program. Analysts, also strewn throughout the government's intelligence community, made estimates that were based mainly on wild speculation about what they assumed the Soviet Union would and could do. In many cases these estimates were based solely on the American and German experiences, and not in any way based on actual information from the Soviet Union. As a result, both military and civilian policymakers were given the impression that the Soviet atomic program did not pose an immediate threat. Thus, a vicious cycle was created: the poor performance of U.S. atomic intelligence meant the faulty estimates of the Soviet nuclear program would continue, thereby slowing any measures to improve the U.S. atomic intelligence system.

This dynamic stands in deep contrast to the German experience. But why? Considering how successfully the United States conducted the atomic intelligence effort against the Germans in the Second World War, why was the U.S. government unable to create an effective atomic intelligence apparatus to monitor Soviet scientific and nuclear capabilities? Put another way, why did the effort against the Soviet Union fail so badly, so completely, in all potential metrics—collection, analysis, and dissemination? How did we get this so wrong?

1

A Reasonable Fear

The U.S. (Mis)Perception of the German Nuclear Program

The idea for the militarization of atomic energy was realized gradually, beginning in the early twentieth century. The New Zealand–born British experimental chemist and physicist Ernest Rutherford and his partner, the British radiochemist Frederick Soddy, sought to build on the discovery of radioactivity by French scientists in the 1890s. In a series of experiments conducted in 1902 and 1903 at McGill University in Montreal, Rutherford and Soddy demonstrated that the energy contained in an atomic reaction was hundreds of thousands, or even a million, times the energy contained in a chemical reaction of the same mass.[1] "These considerations," Soddy wrote of their discovery, "force us to the conclusion that there is associated with the internal structure of the atom an enormous store of energy which, in the majority of cases, remains latent and unknowable."[2] Of course, at the time no one had the slightest idea of how to effect the release of this energy. In fact, most scientists thought the possibility of such a release would be prohibitively difficult to achieve, if not scientifically impossible. Until a more complete understanding of the structure and properties of the atom could be known, atomic energy would remain only a hypothetical construct.

Between 1904 and 1911, Rutherford systematically investigated these structures and properties, culminating in a groundbreaking 1911 discovery that would dramatically shift the scientific paradigm of atomic physics. In a paper he presented to the Manchester Literary and Philosophical Society, Rutherford announced that the universally held belief that the entire mass of the atom, including all of its positive and negative charge, was contained in a single structure was now obsolete. The so-called "plum pudding" model of the atom said that the atom was a viscous mass of positive charge that had electrons interspersed within. The "plums" were the electrons, while the "pudding" was the positively charged soup. Instead, Rutherford explained, the atom had a small, massive nucleus surrounded by a cloud of orbiting electrons, and this nucleus contained nearly all the mass of the atom (over 99.9 percent), and thus nearly all of its energy.

Rutherford's discovery clearly demonstrated that the future of atomic physics rested in the dissection of the nucleus and a complete understanding of its parts. During the 1920s and early 1930s, scientists in Europe and the United States studied the nuclei of atoms of various elements. This enterprise was significantly aided by the discovery of the neutron in 1932 by a student of Rutherford's, the British physicist James Chadwick. Chadwick, who received the Nobel Prize in Physics in 1935 for this discovery, first began to look for the neutron due to an apparent discrepancy between an element's atomic number and its atomic weight. The atomic number is the count of the protons in an element's nucleus (hydrogen has one proton, so its atomic number is 1; silver has forty-seven protons and its atomic number is 47; and so on up and down the periodic table), and atomic weight is a measurement of the mass of an atom (which includes the mass of protons, electrons, and anything else that may be present inside an atom). The problem Chadwick looked to solve was the fact that the atomic number was different, sometimes radically different, from the atomic weight. For example, helium's atomic number is 2, but its atomic weight is 4; oxygen's atomic number is 8, but its atomic weight is 16; uranium's atomic number is 92, but its atomic weight is 238. Electrons contribute very little to the atomic mass. Since nearly all the mass of an atom is contained in the nucleus, then if the nucleus only consisted of protons, what would account for the significant discrepancies between atomic numbers and atomic weights? Chadwick's answer was the neutron, a subatomic particle with the relative mass of a proton, but with no electric charge.

The advantages of this newly discovered subatomic particle were immediately evident to scientists studying atomic physics. Before the neutron, scientists who wished to investigate the nucleus could bombard it with protons or alpha particles (essentially helium atoms) in the hope that this assault would force some kind of physical reaction within the nucleus. The problem with this method is that both the tools used for bombardment (protons and alpha particles) and the nucleus itself are positively charged. This means that in order to penetrate the electrical barrier of the nucleus, the protons or alpha particles would need to be accelerated to very high speeds and would have to contain an enormous amount of energy to successfully enter the nucleus. This process was extremely expensive, and until the later spread of high-energy physics in the late 1930s and 1940s, it was prohibitively difficult for most experimental scientists.[3]

The neutron, having no electrical charge, could enter a nucleus at much lower speeds (about the speed of sound) and with much less expenditure of energy (only the energy of one-fortieth an electron volt), making it an effective and universally available tool for nuclear exploration. The American physicist Isidor Isaac "I. I." Rabi, the 1944 Nobel laureate who worked on radar and the atomic bomb for the United States during the Second World War, described the advantages of the neutron: "When a neutron enters a nucleus, the effects are about as catastrophic as if the moon struck the earth. The nucleus is violently shaken up by the blow, especially if the collision results in the capture of the neutron. A large increase in energy occurs and must be dissipated, and this may happen in a variety of ways, all of them interesting."[4]

Turning an Idea into a Bomb

One prominent scientist who immediately understood the revolutionary consequences of the discovery of the neutron was the Hungarian-born physicist Leo Szilard. After the First World War, Szilard left Hungary in order to study atomic physics under Albert Einstein, Max Planck, and Max von Laue in Berlin. After receiving his doctorate in 1923, Szilard worked as an assistant to Laue and worked on a series of inventions he patented individually or with his collaborative partner, Einstein.[5] In 1933 he moved to London, where he heard of the discovery of the neutron and had his first true revelation about the atomic nucleus. It occurred to him that if scientists could find an element that is split by one neutron and would then emit at least two neutrons, then this element could sustain a nuclear chain reaction. The two neutrons could then hit other nuclei, releasing two more neutrons in the process, and so on. On March 12, 1934, Szilard applied for his first patent on the chain reaction, entitled "Improvements in or Relating to the Transmutation of Chemical Elements."[6] He followed this with two amendments to the patent, dated June 28 and July 4, 1934. It is here that he took the next step: the liberation of energy from a chain reaction. Szilard argued that if he could find an element in which he could create a self-sustaining chain reaction, and if he could assemble this element in a critical mass, then he could, in his words, "produce an explosion."[7]

Despite this huge step, Szilard still did not have answers to the vast majority of questions that scientists would face between this point and the successful creation of the atomic bomb a decade later. In fact, he still did not know what element would be best for producing a self-sustaining chain reaction. It would be the physicist Enrico Fermi's Italian team in Rome that would determine that uranium, more than any other element, was the key to harnessing the untapped energy of the nucleus. Beginning in early 1934, the Fermi team systematically experimented through the elements on the periodic table. As a result of their experiments, Fermi concluded that heavier elements, like uranium, captured neutrons and became heavier isotopes of themselves, and even in some cases transmuted into heavier, entirely different elements. Thus, Fermi argued that uranium, when bombarded with neutrons, became a man-made element, atomic number 93, called a transuranic element. For this discovery, Fermi was awarded the Nobel Prize in 1938.[8]

Fermi's experiments would end up yielding some of the most groundbreaking discoveries in atomic physics, but interestingly not for the reasons anyone would have believed in early 1938 when Fermi's Nobel was awarded. Fermi's conclusion that neutron bombardment led to transuranic elements was proven incorrect in late 1938, when the German team of Otto Hahn, Lise Meitner, and Fritz Strassmann, in an attempt to build on Fermi's discoveries, instead found a far different, and far more momentous, experimental result. As a consequence of their experiments, they argued that uranium did not become a heavier element when it captured a neutron, but instead split into two smaller elements. The Germans had discovered nuclear fission, and they immediately understood the implications.

When a uranium atom splits, perhaps (as one of several possibilities) into a lanthanum atom and a bromine atom (as elements 57 and 35, respectively), the resulting elements' combined atomic weight is not the same as the atomic weight of the original uranium atom. (Lanthanum's atomic weight is 138.91, and bromine's is 79.9, for a total of 218.81. The atomic weight of a uranium atom is, depending on the isotope, 235 or 238.) The missing mass doesn't just vanish into the ether; it is released as energy. While the energy released from a single atom's fission is not significant, it is important to understand that in each gram of uranium, there are 2.5×10^{21} atoms (that's 2,500,000,000,000,000,000,000), and that it would most likely require several kilograms of uranium to achieve the critical mass necessary to create a self-sustaining chain reaction (and a bomb).[9] Even though an atomic bomb would not convert all of its mass into energy (in actuality it would convert only a small percentage), the discovery of fission proved, at least in theory, that immense energy could be released through a nuclear reaction.

News of the German breakthrough, announced on December 17, 1938, spread quickly around the world, and by January 1939 it was the principal topic of conversation in physics faculties at universities throughout the United States. The historian/journalist Richard Rhodes describes the impact of the announcement of fission on American scientists in his book *The Making of the Atomic Bomb*. Rhodes writes that within a week of hearing of the discovery of fission, Robert Oppenheimer had drawn a basic schematic of an atomic bomb in his Berkeley office. Enrico Fermi, who by this time had immigrated to the United States, is revealed to have remarked that a baseball-sized atomic bomb could destroy an urban area the size and density of Manhattan. "A little bomb like that," Fermi said, "and it would disappear."[10]

Leo Szilard, who in 1938 had come to the United States to conduct research at Columbia University in New York, felt it was a matter of great urgency to convey to the U.S. government the implications of this new discovery. Although he was well respected within the scientific community, Szilard knew that he did not have the prestige or the name recognition to convince the government to pay attention. He did, however, know someone who could get his message to the highest levels of the U.S. leadership: his old friend and invention partner Albert Einstein. In the summer of 1939, Szilard convinced Einstein to sign a letter to President Franklin D. Roosevelt, written by Szilard but in Einstein's name, explaining the dangers and opportunities provided by the discovery of fission. In the letter, dated August 2, 1939, Szilard wrote that "the new phenomenon would also lead to the construction of bombs, and it is conceivable—though much less certain—that extremely powerful bombs of a new type may thus be constructed. A single bomb of this type, carried by a boat and exploded in port, might very well destroy the whole port together with some of the surrounding territory." Additionally, Szilard recommended that "in view of this situation you may think it desirable to have some permanent contact maintained between the Administration and the group of physicists working on chain reactions in America. One possible way of achieving this might be for you to entrust with this task a person who has your confidence and who could perhaps serve in an official capacity." This individual might be tasked with coordinating with government agencies and providing funds to university and industrial research laboratories. The letter ended with a warning that the Germans had already begun research and might soon become dangerously ahead of the United States.[11]

The letter was delivered to the president later that month. Yet despite Roosevelt's agreement with the basic implications of the 1939 Szilard/Einstein letter, the U.S. atomic bomb program made little progress in its first three years. Roosevelt's sole action after learning of the German threat was to appoint an advisory committee under the chairmanship of the director of the Bureau of Standards, Lyman Briggs. The Uranium Committee, sometimes called the Briggs Committee, was made up of representatives from the Bureau of Standards and the armed forces. It met occasionally during the subsequent months, consulting with American scientists about the feasibility of both atomic power and atomic weapons. According to Brig. Gen. Leslie Groves, the future head of the Manhattan Project, "On the basis of these discussions, the committee recommended that the Army and Navy make available a modest sum for the purchase of research materials." The first government appropriation for atomic research was only $600 for the purchase of uranium oxide. Most of the work was to be conducted by universities and private institutions, funded by the military and then later, after June 1940, by the newly created National Defense Research Committee (the NDRC was placed under the leadership of engineer Vannevar Bush, and after its creation the Uranium Committee became one of its subcommittees). Groves estimated that more than two years after the letter to Roosevelt (by November 1941), the U.S. government had spent only about $300,000 on projects related to atomic fission research.[12]

In their book detailing the first years of the Atomic Energy Commission, Richard Hewlett and Oscar Anderson Jr. argue that the U.S. atomic bomb program faced serious difficulties from the beginning:

> Fundamentally, the trouble was that the United States was not yet at war. Too many scientists, like Americans in other walks of life, found it unpleasant to turn their thoughts to weapons of mass destruction. They were aware of the possibilities, surely, but they had not placed them in sharp focus. The senior scientists and engineers who prepared the reports that served as the basis for policy decisions either did not learn the essential facts or did not grasp their significance. The American program came to grief on two reefs—a failure of the physicists interested in uranium to point their research toward war and a failure of communication.[13]

In November of 1941, with U.S. entry into the war imminent, Bush decided he needed to press the issue. He reassigned the Uranium Committee to the Office of Scientific Research and Development (OSRD) and established a planning board to study the engineering of facilities for the production of atomic weapons. That same month, the U.S. National Academy of Sciences created a committee to investigate the difficulties associated with an atomic bomb project. On November 27, 1941, the committee sent a report to Roosevelt that detailed the research taking place throughout the country. As a result of the report, on December 6 Roosevelt authorized the creation of the S-1 Committee, which included Bush; Arthur Compton, chair of the Physics Department and the dean of the Division of Physical Sciences at the University of Chicago; Lyman Briggs; the University of California, Berkeley, physicist Ernest Lawrence; and Harvard University's president, the chemist James Conant. Over the next six months, progress was made toward a viable atomic weapons program in American university laboratories, yet by the summer of 1942 no significant mass-scale production had yet occurred (the Manhattan Project under the Army Corps of Engineers would not be created until August 1942). U.S. atomic research in 1942 was still in the basic science, small-scale laboratory research phase.

The progression of the German atomic bomb program was in many respects similar to its American counterpart. The German government created its own committee on uranium, called the Uranverein (or Uranium Club), to study the properties and potential military applications of fission. The Uranverein included such prominent German scientists as Otto Hahn, Werner Heisenberg, and Paul Harteck, as well as some of the government's leading scientific advisers. The Uranverein convinced the Nazi regime to provide funds for research and assigned research projects to universities and research institutions throughout Germany, such as the Kaiser Wilhelm Institutes, the Reich Research Council, and the Reich Ministry of Education.[14]

However, by the summer of 1942, just when the United States was about to accelerate its own atomic production with the creation of the Manhattan Project, the Germans were ending any serious effort to build atomic weapons. In June, the Uranverein had decided that the separation of uranium isotopes was too difficult a project to undertake during wartime, and instead shifted its fission research to the development of nuclear reactors for powering ships.[15] Hitler, knowing nothing about the U.S. and British atomic programs, concluded that atomic weapons would not be available in time to affect the current war. He thus decided to concentrate German financial resources on developing the V-1 and V-2 rockets to attack Great Britain, as he believed these weapons systems could have a more immediate effect.[16] According to the director of the German atomic bomb program, "The total amount which was spent in nuclear research was in the order of 5 million marks or so, not more than that. For the rocket business perhaps 200 or 300 million marks was spent during the war. . . . I would say it was much less than the factor of one tenth; maybe a twentieth or a thirtieth, or less. . . . It has been said that more money was spent on the [U.S. scientific intelligence mission to discover the status of the German atomic bomb program] than was spent on the entire German atomic energy project."[17]

Cause for Concern

Of course, scientists and government officials in the United States had no idea these decisions were being made in Germany. They assumed the Germans were vigorously pursuing an atomic weapons capability. In hindsight, and to many people today, the fear of the Germans in the atomic field might appear irrational. Yet at that moment, in the summer of 1942, there were several factors that made the Americans' concern about the German atomic bomb program reasonable. Each of these, when taken individually, was cause for anxiety within the U.S. scientific community. Taken collectively, however, they produced a sense of near panic, and impelled American scientists and their government to create a remarkably sophisticated and energetic system of scientific intelligence.

To begin with, there was a widely held belief among American scientists that the German atomic bomb program had a significant head start on that of the United States. Arthur Compton addressed this problem: "But at best I do not see how we can catch up with the Germans unless they have overlooked some possibilities that we recognize, or unless our military action should delay them."[18] Estimates varied, from a minimum of six months to a maximum of two or even three years, but a consensus was reached among the U.S. scientific community that the Americans were far behind the Germans in fission research.

Evidence of German activity had begun to arrive in the United States in the summer of 1939. American scientists returning to the United States from Europe that summer reported a coordinated and intensive research program, centered at the Kaiser Wilhelm Institutes in Berlin, focusing on the separation of uranium isotopes. According to the American scientists, a large group of prominent German physicists and chemists were working on mastering the thermal diffusion method of separating the U-235 isotope from the significantly more concentrated U-238. Uranium isotope separation was considered by most American scientists to be the most direct route toward the production of an atomic bomb.[19]

In June 1939, the German physicist Siegfried Flügge published an article in the scientific journal *Die Naturwissenschaften* titled "Can the Energy Contained in the Atomic Nucleus Be Exploited on a Technical Scale?" Flügge discussed the implications of fission and hypothesized that fission research could lead to atomic weapons. This journal was available not only in Germany but in Britain and the United States as well. If anyone had missed the *Die Naturwissenschaften* article, in mid-August 1939 Flügge published a more accessible version of his argument in the widely read German newspaper *Deutsche Allgemeine Zeitung*. Together, these articles clearly indicated a German interest in the atomic bomb, at least within the German scientific community.[20]

Troubling intelligence also came in from the émigré Peter Debye. Debye was a Dutch physicist, a 1936 Nobel laureate, and the former director of the Kaiser Wilhelm Institute for Physics in Berlin (succeeding Albert Einstein). He arrived in the United States in January 1940 and proceeded to tell American scientists that he had left Germany because, according to the American physical chemist Harold Urey, he had been "forced out of his institute and was not allowed to know what was going on in it." However, Debye "did observe that practically every person in Germany who knew anything about atomic physics or the separation of isotopes, went in and out of his institute."[21]

A final source of information about German science and the Germans' interest in atomic fission that would influence American scientists in summer 1942 arrived through what came to be known as the "scientific underground." This term refers to a loose consortium of European scientists who had remained in Europe during the war, opposed the Nazi regime, and covertly passed along information about German scientists to their American, British, or émigré colleagues. Leo Szilard provided some information from the scientific underground. He explained that in September 1941, his friend the physicist John Marshall had given him startling news from Germany. Szilard wrote that "a son of [the German physicist Friedrich] Dessauer who arrived from Switzerland told Marshall that according to his information the Germans got a chain reaction going."[22] If accurate, this was alarming since it would still be more than a year before the Americans would achieve their own self-sustaining chain reaction. To Szilard and many of his American colleagues, this indicated that the Germans were significantly ahead.

Germany also had (or had acquired through conquest) all the necessary facilities, materials, and industrial infrastructure needed to initiate a successful atomic bomb project. No one in the United States doubted the might of German industry. If the Germans had dedicated a significant proportion of their industrial resources to an atomic bomb project, American scientists were convinced they would be successful. Germany was also the home of many of the world's greatest scientific laboratories, including the Kaiser Wilhelm Institutes for Physics and Chemistry but also the university laboratories at the Universities of Gottingen, Leipzig, Cologne, Hamburg, Giessen, Heidelberg, and Vienna (through the Anschluss). Each of these laboratories maintained state-of-the-art facilities for fission research, and thus were potential contributors to a German atomic bomb project. Additionally, the German military's advance through Europe had given German scientists the use of what were arguably the most advanced laboratories on the Continent: Niels Bohr's institute in Copenhagen and Frédéric Joliot-Curie's laboratory at the Collège de France in Paris. Each of these laboratories had provided the Germans with a key piece of experimental technology that they were previously lacking, and that was essential for any serious atomic bomb program: the cyclotron.

The cyclotron was invented in 1932 by the American physicist Ernest Lawrence, eventually earning him the Nobel Prize in 1939. The machine accelerated charged subatomic particles (protons or alpha particles) to high speeds that enabled them to penetrate the nucleus of an atom and alter its atomic structure (hence its alternate names: particle accelerator and atom smasher). The cyclotron had many uses, the most important of which was that it allowed scientists to study nuclear reactions in an entirely novel way. By providing the means to bombard nuclei with positively charged particles at high speeds and with high energy, the cyclotron significantly lowered the learning curve for discovering the secrets of atomic weapons. Under the right circumstances, the cyclotron could even create fissionable artificial materials such as plutonium, otherwise inaccessible to the Germans without the possession of a working nuclear reactor.

Of course, uranium is necessary to produce plutonium, and the Germans had as much of this element as they would need for any atomic bomb program. They acquired much of their uranium through the German conquest of Czechoslovakia, which brought them the most productive uranium mine in Europe at Joachimsthal, in the Ore Mountains of what is today the Czech Republic. It was at Joachimsthal that uranium was first discovered by Martin Heinrich Klaproth in 1789 (he named it as well). Later, the Joachimsthal mines provided Marie and Pierre Curie with the materials from which they would ultimately separate radium and polonium.[23] But while Joachimsthal was a key source of uranium, it paled in comparison to what Leslie Groves argued was the "most important source of uranium ore during the war years," the Shinkolobwe Mine in the Belgian Congo.[24] The Belgian firm Union Miniére had been mining Shinkolobwe throughout the 1930s, and had shipped thousands of tons of uranium ore to Belgium. When the Germans invaded Belgium in 1940, some of that ore was smuggled out to Britain and the United States, but hundreds if not thousands of tons of uranium ore were assumed to have been captured by the Germans.[25]

The Germans had also acquired through conquest a hydroelectric plant in Norway that produced a large quantity of a substance called "heavy water" (or D_2O; *D* is deuterium, an isotope of hydrogen with an atomic mass of 2 since it has a single neutron, thus "heavy"). Heavy water could be used as a moderator in a nuclear reaction. A moderator is a substance used to slow down the neutrons emitted in a fission reaction. If neutrons are moving too fast, they will not interact with the fissionable material in a way sufficient to produce a self-sustaining chain reaction. Slowing down a neutron gives it a better chance to interact with the particles inside the nucleus, thereby statistically increasing the chance of a successful reaction. The atoms of heavy water, or any other moderator such as graphite or beryllium, slow neutrons by bumping into them and decreasing their velocity, thus increasing the probability for neutron absorption and, in a practical sense, reducing the amount of material needed to reach critical mass.[26]

The U.S. bomb program would eventually choose graphite as its moderator, but initially it entertained the use of a heavy water moderator. For many scientists in the United States, the advantages of heavy water as a moderator were obvious. However, they believed that the time it would take to build the industrial infrastructure for producing an adequate supply of heavy water for an atomic bomb program was prohibitive.

This would certainly have been true for the Germans as well. But instead of having to build their own heavy water plant, in 1940 the Germans captured the Norsk Hydro plant in Rjukan, Norway, the only operational heavy water facility in Europe. At full capacity, the Norsk Hydro plant could easily produce enough heavy water for an aggressive atomic bomb program. In December 1941, Harold Urey brought news back from British sources that the Germans were manufacturing a substance called "heavy paraffin" from heavy hydrogen produced in Norway. According to Arthur Compton, "This could only be for a moderator for a fission reactor."[27]

The Aura of German Science

Equally important to the acquisition of materials was the American perception of German scientists. German science had long been considered the best in the world, and German scientists were revered by many of their American counterparts as paragons of creative scientific accomplishment. The Kaiser Wilhelm Institutes in Berlin were the centers of worldwide physics and chemistry in the 1920s and 1930s. At one point in the 1920s, the directors of the various Kaiser Wilhelm Institutes included a roster of scientific giants: Albert Einstein (Physics), Fritz Haber (Physical Chemistry), and the future discoverer of fission, Otto Hahn (Chemistry). A graduate student at any of the prominent German institutions could expect to receive instruction from such leading figures as Einstein, Max Planck, Erwin Schrödinger, Max Born, Paul Ehrenfest, Arnold Sommerfeld, and Max von Laue. During the first half of the twentieth century, Germany won more Nobel Prizes in science than any other nation, most of them in physics and chemistry. Many of the leading Manhattan Project scientists had done their graduate or postdoctoral work in Germany, including Robert Oppenheimer, Hans Bethe, Edward Teller, Enrico Fermi, Wolfgang Pauli, and Victor Weisskopf, to name only a few of those working on the U.S. atomic bomb.

Despite the mass emigration of Jewish scientists from Germany to Britain and the United States in the 1930s following Hitler's rise to power, many outstanding scientists still remained in Germany, including several of the world's most exceptional minds. The most important German scientist was undoubtedly Werner Heisenberg. After the war, Samuel Goudsmit, a Dutch-born American physicist and the future scientific chief of the U.S. scientific intelligence mission to Europe in 1944–45 (see chapters 3 and 4), would describe the American scientists' perception of Heisenberg when the Second World War began: "To an outsider, a professor is a professor, but we knew that no one but Professor Heisenberg could be the brains of a German uranium project and every physicist throughout the world knew that."[28] Later in his book, Goudsmit further clarified his feelings about Heisenberg when he wrote, "He is still the greatest German theoretical physicist and among the greatest in the world. His contributions to modern physics rank with those of Einstein."[29]

Heisenberg was a prodigy. He was only twenty-two when he earned his PhD in physics under Arnold Sommerfeld in Munich. By twenty-six he was a professor of theoretical physics at the University of Leipzig, and by thirty-two he was a Nobel laureate for his work on quantum theory. By the time of his fortieth birthday, he had been appointed the director of the Kaiser Wilhelm Institute for Physics, the premier physics facility in Germany and perhaps worldwide, and what American scientists assumed would be the center of German atomic bomb research.

Heisenberg's meteoric rise to prominence began when his genius was recognized at an early stage in his scientific career. Sommerfeld was quick to identify Heisenberg's potential, particularly when the young German physicist was able in his first semester to work through a physics problem that had been unsolvable to physicists with two or three times his experience and education. In June of 1922, Sommerfeld took his prized student to meet Niels Bohr in Göttingen, during an annual series of talks given by the founder of atomic theory. These lectures were frequently attended by notable physicists throughout the world, and Bohr and the rest of the elite of the European scientific community were amazed when a young German graduate student directly questioned aspects of Bohr's theories. Some professors were offended by Heisenberg's presumption, but Bohr was not. According to Amir Aczel, "He had found a keen mind, a young person who could really understand quantum theory in depth, at a level beyond that of anyone else in the audience." From this moment, Bohr and Heisenberg would become close friends and scientific collaborators.[30]

After Heisenberg received his doctorate, he left Munich in 1924 to join Max Born in Göttingen. Born, a close associate of Einstein's and one of the founders of quantum mechanics, like Sommerfeld saw the promise in Heisenberg, and the two of them, along with the contributor Pascual Jordan, developed the matrix mechanics formulation of quantum mechanics in 1925, along with a number of other papers on quantum mechanics that furthered the development of quantum theory.[31] A year later, Heisenberg moved to Copenhagen to be Bohr's assistant at the Institute of Theoretical Physics and a lecturer at the University of Copenhagen. It was in Copenhagen that Heisenberg developed the principle that he is most known for today. In 1927 he proposed the idea that a particle's momentum and its position could not be known simultaneously with precision—the so-called Uncertainty Principle. It stated that the more precisely a particle's position is known, the less precisely its momentum can be calculated, and vice versa.[32] Heisenberg's theories revolutionized physics and would eventually allow for the development of modern electronics, including most of the computer-based products used today.[33]

Heisenberg's talents in theoretical physics were well appreciated by American and émigré scientists, but what truly worried the U.S. scientific community were events that took place in the summer of 1939. Just as tensions in Europe were reaching their height, and as war seemed imminent, Heisenberg traveled across the United States, meeting with many of his old colleagues. Some had immigrated to the United States from Europe, while others were Americans who had studied in Europe during the 1920s and 1930s. At each stop, Heisenberg was urged to leave Germany and Hitler behind and take a teaching and researching position in the United States.

At the University of Rochester in New York, Victor Weisskopf and Hans Bethe (who had come to Rochester from Cornell to speak with Heisenberg) prevailed upon Heisenberg to take a job in the United States. They told him that he could essentially choose where he wanted to live and teach, and that any U.S. institution would bend over backward to open a faculty position for the brilliant German physicist. Eugene Wigner urged Heisenberg to take a job at Princeton, while the physicists I. I. Rabi and George Pegram at Columbia University again extended an offer (Pegram had first offered Heisenberg a job in 1937). He spoke with a group of scientists at the University of Chicago and with Robert Oppenheimer at Berkeley, where he was met with similar requests and provided similar refusals. At the end of July, Heisenberg stayed for a week in the home of his friend Samuel Goudsmit, whom he had known since 1925, at the University of Michigan in Ann Arbor. There he spoke at length with Goudsmit and Enrico Fermi, who had also known Heisenberg since the 1920s when they had met in Göttingen. In Michigan, both Goudsmit and Fermi expressed their belief that the United States offered the best opportunity for Heisenberg to continue his groundbreaking research. Once again Heisenberg politely refused.

What worried American scientists the most were Heisenberg's reasons for returning to Germany. After speaking to him, most of the American scientists (and European émigrés) came to the same conclusion as to Heisenberg's motives: he was a German nationalist at heart. His country needed him. Despite the ominous nature of the Hitler regime, Heisenberg would stay in Germany to ensure that he could do whatever it took to help his country in the coming war. The greatest fear for scientists in the United States was that this would include the development of a German atomic bomb.[34]

If Heisenberg had indeed decided to stay in the United States, the most dangerous scientist in Germany would have become the chemist Otto Hahn. According to the chemist Glenn Seaborg, the discoverer of plutonium and a participant in the U.S. bomb program, "Hahn was the undisputed world leader in radiochemistry; his book *Applied Radiochemistry* [published in 1936] was my bible."[35] As described above, in 1938 Hahn, along with Lise Meitner and Fritz Strassmann, was the first to discover nuclear fission after bombarding uranium with neutrons. On December 22, 1938, Hahn sent the results to the journal *Die Naturwissenschaften*, which announced to the world the revolutionary findings. In February of 1939, Hahn and Strassmann published a second article in *Die Naturwissenschaften* in which they predicted the release of additional neutrons during the fission process. Later proven correct by the French scientist Frédéric Joliot-Curie, Hahn's prediction served as the basis for the concept of a chain reaction, and ultimately for the atomic bomb.

By this time, Hahn had already established himself as a giant in the field of chemistry. At the turn of the twentieth century, he had discovered several isotopes of thorium (radiothorium, mesothorium 1 and 2, and ionium) and had been nominated for the Nobel Prize for his discovery of mesothorium 1. In 1912 Hahn became head of the Radioactivity Department of the Kaiser Wilhelm Institute for Chemistry, and from 1928 to 1946 he was director of the Kaiser Wilhelm Institute for Chemistry. In 1924, Hahn was elected a full member of the Prussian Academy of Sciences in Berlin. His nominators for this prestigious post included Albert Einstein, Max Planck, Fritz Haber, and Max von Laue. In between his early discoveries and the recognition of his accomplishments, he served in the German Army during the First World War as a chemical warfare specialist under the command of Fritz Haber.[36] Hahn was eager to go to war and serve his country.[37] As a member of Haber's Pioneer Regiment, Hahn participated in poison gas experiments and attacks on both fronts, and he strove to make the gas as efficient and deadly as possible. This history made it probable that Hahn would do what was necessary to assist the German government during the Second World War in its quest for atomic weapons. He had not hesitated to help develop weapons of mass destruction in the past, and the American scientists assumed he would support the Nazis in the present.[38]

Heisenberg and Hahn were the two most famous of the scientists who remained in Germany during the Second World War, but by no means were they the only scientists of considerable skill available to the Nazi atomic bomb project. Another scientist of note was Paul Harteck, an expert in isotope separation and a specialist in heavy water production. Harteck was a physical chemist who had earned his PhD at the University of Vienna under Max Planck. In 1928 he became the primary assistant to Fritz Haber at the Kaiser Wilhelm Institute for Physical Chemistry, and in 1933 he won the Rockefeller Fellowship to study with Ernest Rutherford at the Cavendish Laboratory in Cambridge, England. While in Cambridge, he, Rutherford, and Marcus Oliphant had jointly discovered the hydrogen fusion reaction (a key discovery for the later development of thermonuclear, or hydrogen, bombs). In 1934, Harteck was named the director of the Institute for Physical Chemistry in Hamburg, a post he held until 1951.[39] His expertise in isotope separation, and in particular his experience working with heavy water, made him a natural and, in the minds of the Americans, incredibly dangerous choice for the German atomic bomb program.

Other prospective German atomic bomb scientists included Hans Geiger, Wolfgang Gentner, Pascual Jordan, Klaus Clusius, Walther Gerlach, Walther Bothe, Erich Bagge, Max von Laue, Fritz Strassmann, and Karl Wirtz. Geiger, who had worked with Ernest Rutherford, was the inventor of the device that bears his name, used for detecting ionizing radiation. Gentner was an able experimental physicist who had worked with both Ernest Lawrence in the United States and Frédéric Joliot-Curie in France. He was an expert in cyclotron operations and would be a key asset for the Germans if they wanted to construct their own atom-smashing machines. Jordan, as explained above, worked with Heisenberg and Born on the conceptualization of quantum theory. In 1933 he became a member of the Nazi Party and later that year joined a Brownshirts (Sturmabteilung) unit.[40]

Klaus Clusius was the director of the Physical Chemistry Institute of the University of Munich in the 1930s. There he conducted major experiments on heavy water and developed, along with a colleague, the thermal diffusion method of isotope separation. Walther Gerlach was an internationally known physicist who had done groundbreaking work in the early 1920s (he codiscovered a phenomenon known as the Stern-Gerlach effect).[41] He was known in the United States and Britain to have connections with the Gestapo.[42]

Erich Bagge, a member of the Kaiser Wilhelm Institute for Physics, was also a specialist in isotope separation, while Max von Laue was a Nobel laureate for his discovery of X-ray diffraction in 1914. Bothe, Strassmann, and Wirtz were all world-class experimentalists. Bothe worked at the Kaiser Wilhelm Institute for Medical Research, Strassmann was a colleague of Otto Hahn's who had helped Hahn demonstrate the unknown phenomenon of nuclear fission, and Karl Wirtz was a good friend of Heisenberg's. He was a close collaborator and made up for what was, perhaps, Heisenberg's only significant weakness: his lack of experimental experience. Wirtz knew how to run a lab and how to create and manage large-scale experiments. He was considered Heisenberg's right-hand man.

The Americans also worried about several non-Germans' potential contributions to the German atomic bomb program. Two Italians, Edoardo Amaldi and Gian Carlo Wick of the University of Rome, had been colleagues of Enrico Fermi's before the war. They were both excellent physicists who had worked with Fermi in his experiments with radioactivity (for which Fermi won the Nobel Prize in 1938). As Amaldi and Wick were now citizens of an Axis nation, the Americans worried that they would be pressured to work for the German war effort. The most effective use of their skills, Fermi and others feared, would be the German atomic bomb project.

Finally, two French physicists, the married couple Frédéric and Iréne Joliot-Curie, could provide considerable assistance to the Germans. Both Iréne and Frédéric were outstanding physicists in their own right. Iréne was the daughter of Pierre and Marie Curie and followed in her parents' footsteps when she received her PhD in physics in 1925 from the Sorbonne. Frédéric was an assistant to Marie when he met and fell in love with her daughter, whom he married the following year. The two decided to work together and in January 1934 announced that they had been able to induce artificial radioactivity, a feat that would lead directly to Hahn's discovery of fission in 1938. In 1935, they had jointly been awarded the Nobel Prize in Chemistry for this discovery, and by the time the war began they were continuing their research at the Collège de France in Paris, where Frédéric had built a cyclotron and was working on building a chain-reacting nuclear pile (Iréne was diagnosed with tuberculosis and was not able to work for several years). Along with Heisenberg and Hahn, Frédéric Joliot-Curie was the most prominent and decorated physicist remaining in Europe.

It was also widely believed that the German atomic bomb program had all the necessary support from the Nazi political hierarchy for an ambitious research effort. There was specific evidence that the German government was interested in fission research. The United States had learned that the Germans had suspended the sale and export of uranium from their mines in Czechoslovakia. To the Americans, this indicated that the Germans were hoarding the material in an effort to acquire an adequate supply for bomb research. This information was supplied to the U.S. government via the Szilard/Einstein letter to President Roosevelt in 1939.[43] American scientists had also heard of a story related to Harold Urey in the summer of 1940 by a Colonel Zoring of the U.S. Army's Ordinance Department. Zoring was attached to the German Army as an official observer during its invasion of France, and described an occasion when a German officer went in search of French physicists to recruit for German war research. The German officer explained that all the German physicists were busy in Germany working for the regime on atomic research.[44]

American scientists also knew of a number of connections between German science and high-level officials within the German government. Reichsmarschall Hermann Göring, Hitler's designated successor and chief deputy, was the titular head of the Reich Research Council, the main German agency coordinating atomic fission research. Heinrich Himmler, the commander of the German Home Army and the man most responsible for the policies behind the Holocaust (other than Hitler, of course), was an old family friend of Werner Heisenberg's. This fact was widely known to the Americans and they feared this relationship would be exploited to promote atomic research to the German High Command.[45] The most direct relationship between German science and the German government was that of the German physicist Carl Friedrich von Weizsäcker, a colleague and close friend of Heisenberg's. An experienced and capable physicist in his own right, Weizsäcker was assumed to be an integral member of any German atomic bomb program. This is not, however, why he was notable to the Americans, or why he is the only scientist specifically mentioned in the Szilard/Einstein letter to Roosevelt. Weizsäcker's father was Ernest von Weizsäcker, one of Hitler's top diplomats and the man who would become state secretary in Joachim von Ribbentrop's Foreign Ministry. Carl's family was notable in other ways: his grandfather was the last prime minister of the Kingdom of Wurttemberg, and his brother would become mayor of West Berlin in the early 1980s, president of the Federal Republic of Germany (West Germany) in 1984, and then the first president of the reunited Germany.

Close ties to high-ranking members of the German government such as these could guarantee that atomic research was taken seriously by those most capable of providing the necessary funding and support. A major concern among American scientists was the question of whether Hitler and the Nazi regime would authorize an all-out effort in the nuclear field. If they did, many within the U.S. scientific community believed that an authoritarian government like that of Germany would be a better sponsor for atomic research than the government of the United States, because, as Samuel Goudsmit put it, "totalitarianism gets things done where democracy fumbles along, and that certainly in those branches of science contributing directly to the war effort the Nazis were able to cut all corners and proceed with ruthless and matchless efficiency."[46] According to Leslie Groves, American scientists were "aware of the pressures that certainly would be brought to bear on the German scientists to ensure their utmost support of their country's military program."[47] If this pressure was successful, and German science beat the Americans to the bomb, the nightmare scenario could come true—Hitler with an atomic bomb. No one on the Allied side, from the scientists to the military to the British to the U.S. civilian government, doubted that Hitler would waste any time using his new technological wonder weapon in a devastating attack on the Allies.[48]

Even if Germany was unsuccessful in completing a working atomic bomb, however, there was a reasonable fear among American scientists that the German fission program could produce enough radioactive material to create an offensive weapon for spreading deadly radioactivity on London or on concentrated troop formations. Arthur Compton was particularly fearful of this eventuality. He had first thought of the potential of radioactive weapons as early as 1941 (when he suggested developing them for American use).[49] In the summer of 1942, he was principally concerned with Allied vulnerability to attack by German radioactive bombs. In a memorandum entitled "Protection Against Ionizing Bombs," Compton urged the chairman of the National Defense Research Board, James Conant, to take action to protect likely Allied targets. According to Compton, American scientists had "become convinced that there is real danger of bombardment by the Germans within the next few months using bombs designed to spread radio-active material in lethal quantities." Compton was most worried about the vulnerability of British cities and industry:

> You will probably have learned from Mr. Bush that apparently reliable information has reached us to the effect that the Germans have succeeded in making the chain reaction work. Our rough guess is that they may have had the reaction operating for two or three months. When they reach the hundred thousand kilowatt stage in their power plant, they will be producing radio-active material fast enough to supply bombs of about 100,000 Curies each, daily. Exploded inside important industrial plants, these would make them uninhabitable for some months (half life about two weeks). We anticipate that our experimental plant will be producing such radio-active materials in amounts of military importance before the end of this year. The Germans may have them already.[50]

There was also concern that the Germans could use radioactive weapons to attack large groups of troops on the battlefield, or perhaps troops gathered at an embarkation point. An obvious target would be the Allied soldiers who would later mass in British ports for the invasion of Normandy. The Germans could drop radioactive material on those troops or could even irradiate the English Channel in order to prevent its crossing.[51] Or they could drop radiation bombs on the Allied soldiers once they had established the Normandy beachhead, a fear so acute that some American officers during D-Day were equipped with Geiger counters and U.S. Army doctors were warned to be on the lookout for any signs of radiation poisoning.[52]

A less likely scenario, but one still feared by American scientists, would be a German radiological attack on cities within the United States. The difficulties the Germans would have getting the weapons across the Atlantic may not have been sufficiently considered by the American scientists, but they worried that potential targets for the Germans could be the water and food supplies of major U.S. cities. The Germans, Goudsmit wrote, could use "chemically non-detectable substances and sow death wholesale among us by dreadful invisible radiations." Goudsmit described the sense of apprehension within the U.S. scientific community:

> The fear was so real that scientists were even sure of the place and the date of Hitler's supposed radioactive attack. The Germans must know, they thought, that Chicago was at that time the heart of our atom bomb research. Hitler, loving dramatic action, would choose Christmas day [1942] to drop radioactive materials on that city. Some of the men on the project were so worried they sent their families to the country. The military authorities were informed and the fear spread. I heard rumors that scientific instruments were set up around Chicago to detect the radioactivity if and when the Germans attacked.[53]

The final reason for the United States' concern had less to do with German atomic progress than with atomic research in the United States. All of the above reasons for American concerns about the German atomic bomb program were not new realities in the summer of 1942. The question then must be asked: why then? Why in the summer of 1942 did it become so vital for the United States to learn what the Germans were up to? Why not earlier, when so much of the information was first learned? Why not later?

The answer lies in the considerable progress that American scientists made in the first months of 1942. When the idea of an atomic bomb was simply that, an idea, a theoretical construct relegated to the chalkboards of university laboratories, there was less to fear from a German atomic bomb program. If the bomb could not be built, then it also could not be built by the Germans, regardless of how scientifically talented they might be. But when atomic weapons changed from an intellectual exercise in theoretical physics into a very likely reality, the American scientists' fear of a German atomic bomb increased dramatically. In the six months leading up to the correspondences of June 1942 outlined in the beginning of this chapter, the U.S. atomic bomb program reached major theoretical milestones that made the construction of a nuclear weapon seem much more achievable. The American scientists celebrated their discoveries, but the victory was bittersweet: if it could be done by the Americans, it could be done by the Germans.

In the winter of 1941–42, the scientists at Ernest Lawrence's Berkeley laboratory made considerable progress on the separation of U-235 and concluded that they could replicate their success on the scale necessary for mass production. By the spring, it looked as though the progress made by American scientists could shorten, by perhaps as much as six months, the previously estimated time before enough material would be available for strategic use (to make a working bomb). Scientists in the spring of 1942 also saw new reasons to be hopeful about the power and efficiency of atomic weapons. The size of the necessary critical mass was calculated to be much smaller than what was expected just months earlier, due in part to new discoveries regarding the fissionability of U-235—it was found to be much more fissionable than earlier believed, particularly in how it reacted to "fast" neutrons. Furthermore, the Americans were by that time convinced that they had seriously underestimated the destructive force of an atomic bomb. New calculations suggested that the yield of an atomic bomb would be at least three times the predicted yield from six months earlier (an estimated 2,000 t yield in spring/summer 1942 versus 600 t in late 1941).[54]

By late April 1942, all of the pieces were in place for the U.S. atomic bomb program to take the next step out of the university laboratory and into full-scale, government-sponsored, and government-run production. Arthur Compton presented the complete argument to the S-1 Committee. He argued that a chain-reacting pile (a nuclear reactor) was feasible and imminent, that the processes for U-235 isotope separation were working better than expected, that Glenn Seaborg had demonstrated an efficient method for separating plutonium chemically from uranium, and that the design for a mass-scale plutonium production plant was soon to be realized.[55] By the summer, engineering studies had shown that plutonium, like U-235, could be produced in quantity. As a result of these conclusions, the task of producing both of these fissionable elements would be assigned in June 1942 to the U.S. Army Corps of Engineers, who began initial construction in August.

Despite the forward movement of the U.S. program, and, as has been demonstrated here, *because* of the progress of the U.S. program, scientists in the United States were still terrified by the possibility of a German atomic bomb. Samuel Goudsmit expressed what most American scientists were feeling: "Our scientists realized clearly the dreadful implications of the atom bomb, if it could be put together, and being men of good will many of them secretly hoped that the thing would be too difficult to achieve during the war. When they found out it was not only not impossible, but highly probable, that they could make an atom bomb that would work, they became a little scared, more than a little. The thought of German superiority drove them almost to panic."[56]

Stealing the March

The scientists in the United States working on nuclear physics and chemistry were universally convinced of the danger Germany represented. The only thing left to do at that point was to convince the U.S. government to take the problem as seriously as they did. Thus on June 1, 1942, Leo Szilard wrote a letter to Arthur Compton urging the government to begin a concerted effort to discover the status of the German atomic bomb program.[57] Compton agreed. Unlike Szilard, however, Compton had the prestige and authority to take action. He had won the 1927 Nobel Prize in Physics for the discovery of the "Compton effect," which describes the scattering of a photon of light in an interaction with an electron—evidence of light's duality as both a wave and a particle. Also, in addition to his duties at the University of Chicago, Compton was the head of the OSRD's S-1 Committee, which was tasked with investigating the properties and manufacture of uranium for potential use in atomic weapons. According to Glenn Seaborg, Compton's role in S-1 was to supervise the early design of the atomic bomb, and "until the War Department took control with the Manhattan Project in the fall of 1942, Compton was the de facto leader" of the U.S. program.[58]

Compton wrote his own letter on June 22, and sent it straight to the top of the U.S. scientific hierarchy: OSRD chair Vannevar Bush. Compton told Bush that it was essential that the United States do something to gather information about the German atomic program, and warned him that he had "recently become aware that the threat of German fission bombs is even more imminent than we supposed [a month earlier]." He continued by outlining the lack of current options the Americans believed they had: "Secret service activity in German is urgently needed, to locate and disrupt their activities. Perhaps our physicists can give helpful advice to this end. Our careful consideration of possible counter-measures has led to nothing except such destruction at the source and blocking of planes, etc., bringing their bombs to us." Compton's most ominous warning concerned a potential timetable for German atomic capability: "If the Germans know what we know [about the production of plutonium]—and we dare not discount their knowledge—they should be dropping fission bombs on us in 1943, a year before our bombs are planned to be ready."[59]

Compton's letter had the desired effect. Bush was convinced of the importance of the creation of a viable, effective scientific intelligence program targeted at the German atomic bomb program. The only question that remained in the summer of 1942: how to begin?

2

Making Something out of Nothing

The Creation of U.S. Nuclear Intelligence

During the summer of 1942 the U.S. scientific leadership made two key national security decisions. In both cases, these choices were prompted by an acute fear of the dire threat posed by German atomic bomb research. The first, initiated in mid-June by the OSRD chair and presidential science adviser Vannevar Bush, was the decision to shift the U.S. atomic bomb project from the experimental stage to development and production. On June 17, Bush asked for and received permission from President Roosevelt to begin this transformation, and by the end of the summer a crash program to build an atomic bomb was formed under the leadership of Brig. Gen. Leslie Groves of the Army Corps of Engineers. Groves chose the University of California, Berkeley, physicist J. Robert Oppenheimer to be the scientific chief of the Manhattan Engineer District (MED), informally known as the Manhattan Project. Over the next three years, Groves and Oppenheimer would together shape and guide U.S. bomb research and manufacture to their successful conclusion.[1]

The second key decision was concerned with scientific intelligence. Before the summer of 1942 the United States had never paid attention to an enemy nation's basic laboratory research and scientific discovery. Instead the emphasis had been placed on technological intelligence, or the practical applications of scientific research or engineering in the form of the production and deployment of weapons systems or related military equipment. Technological intelligence had been a mainstay of U.S. national security since the time of the American Revolution,[2] and the Second World War was no exception. Captured Enemy Material Units from the Economic Intelligence Division of the Office of Economic Warfare were deployed to each theater of operation. Once there, they sent back to Washington reports of enemy weapons and equipment, including detailed technical information on radar, aircraft, engines, armament, chemical and biological weapons equipment (such as protective masks), munitions, armor, petroleum products, and gear such as rangefinders and medical equipment.[3]

Scientific intelligence presented unique challenges not normally faced in other, more traditional types of intelligence. The majority of intelligence officers in the 1940s did not possess a scientific or technical education. Instead, most intelligence professionals were trained to evaluate political, military, or economic data and determine what was, or was not, actionable intelligence. This meant that even if they were able to infiltrate the scientific establishment of Germany, average intelligence officers engaging in human intelligence collection (or HUMINT) would not have the necessary background to understand the kinds of information they should be seeking. To complicate matters further, science comprises multiple distinct fields, each of which requires specialized knowledge. There is biology, physics, chemistry, geology, astrophysics, and, of course, nuclear physics. In other words, just because the government employed a highly skilled "scientist" does not mean that he or she had the necessary educational background to understand scientific intelligence outside his or her particular field. The U.S. government could not treat science as it did economics or politics (and hire a generic economist or a political scientist), but instead would be forced to recruit scientists from all security-related fields (that is, essentially all scientific fields) in order to assess German capabilities. Furthermore, scientific language presented serious problems for those tasked to correctly translate scientific intelligence into English. Even the best-trained linguist may not have the skills necessary to translate highly technical data from German to English, and certainly not in the timely manner required by the pace of the war. In essence, what was needed was a highly educated scientist with excellent linguistic capabilities. Unfortunately, this skill set was very rare.

From the Drawing Board

Thus the U.S. scientific leadership was tasked not only with discovering the extent to which German atomic science had progressed, but also with designing and developing, without any historical precedent, the scientific intelligence organization that would accomplish this mission. The American scientists' efforts led to mixed results. To some extent, scientists were well qualified to investigate some of the major problems outlined in chapter 1—specifically, the potential threat of radiological attack on the United States. Arthur Compton, who was especially concerned about this eventuality, assigned J. C. Sterns, one of his physicists at the University of Chicago's Metallurgical Laboratory, to investigate possible defenses against such an attack. Sterns was chosen for this task not only because Compton considered him "one of our most capable men," but also because any such project would require close collaboration with the military, and Sterns was "suitable for cooperation with the army in connection with the use as well as the development of the appropriate detection devices."[4] In a letter dated July 16, 1942, Compton told Sterns that a German radiological attack on the Allies was likely and could be a real possibility before the end of the year. He argued that the United States must "take urgent steps to prepare" for this contingency, and ordered Sterns to "free yourself of your present duties at the earliest moment possible and . . . accept the assignment of organizing and carrying through a program designed to build up a defense against" radiological attack.[5]

American scientists were also able to tap into their contacts in Europe, the so-called scientific underground, for information about German atomic research. Scientists in Europe passed along information about the location and activities of German scientists, including Heisenberg, Hahn, and Weizsäcker, as well as Frédéric Joliot-Curie.[6] However, in many cases the reports were ambiguous, and they sometimes conflicted with one another as well as with what the Americans understood to be the truth. In some cases the information came from state-controlled German newspapers, or even from third- or fourth-hand reports that had worked their way through many different people (and perhaps many different revisions) before arriving in the United States. With scientific information this is a particularly vexing problem. Nuanced mistakes in scientific reporting, or minor changes to information via transmission, can have a major impact on the veracity of the information. For example, in the letter from Leo Szilard to Arthur Compton detailed in chapter 1, Szilard argued that the Germans might already have achieved a self-sustaining chain reaction. He based this contention on information he had received from his friend John Marshall in Switzerland, who in turn had heard this "fact" from a son of the German physicist Friedrich Dessauer, who himself had learned it from his father.[7] The result of this scientific game of telephone, in many cases, was a warped, inaccurate picture of the scientific situation in Europe, and it thwarted any precise assessment of the reality of the progress of the German atomic bomb program.

Another approach to the intelligence problem was to make estimates of German progress based on scientific expertise and knowledge of U.S. atomic bomb research. Harold Urey, a Nobel Prize winner in 1934 for his discovery of heavy water, provided Bush with his best guess as to how much success the Germans had had with the use of heavy water in their isotope separation effort. He argued that while it was impossible to know for certain the progress made by the Germans, "I think it is unsafe to assume that they were less efficient in their development than we have been." Urey continued, "It is therefore reasonable to assume that they had a one year start on us and that [since] it will require two years for us to get a plant in operation [for full-scale isotope separation] we should assume that by a year from now [the summer of 1943] the Germans will have such a plant going."[8] The Hungarian American theoretical physicist and mathematician Eugene Wigner would also contribute to this effort. Wigner, who would win his own Nobel Prize in Physics in 1963, used his knowledge of the American uranium isotope separation and plutonium production program to create a memorandum that presupposed a German schedule for the production of fissile material. Like Urey (and many other American scientists) he concluded that the Germans would have enough uranium or plutonium by the summer of 1943 to make at least one, but more likely several, atomic bombs.[9]

The most coordinated effort to use American scientific expertise to estimate the extent of the German atomic bomb program was directed by Bush and the OSRD. Using his knowledge of the U.S. program, Bush tried to create a detailed analysis of the characteristics that could be found in any plant designed to manufacture fissile material for use in an atomic bomb. Perhaps distinctive elements such as water supply, the temperature of cooling water, and the availability of electrical power sources could be utilized to identify atomic research facilities through airborne surveillance.[10] Using information about German infrastructure he obtained from the Board of Economic Warfare, Bush hoped that he could locate German facilities that could be targeted and destroyed by U.S. and Allied bombing.

Unfortunately, this expedient proved worthless. When Bush first proposed this approach in early July 1942, he believed it obligatory to pursue this method of analyzing potential clues about the German program even though he also felt that it was "none too promising." He resigned himself to "follow them up" wherever they might lead.[11] But by late September, Bush would reluctantly accept the fact that the attempt to determine the location of German fissile material production plants by studying water supply and available power sources "does not seem to get anywhere."[12] In a memorandum provided to Army Intelligence chief Maj. Gen. George V. Strong, Bush detailed the reasons for failure, which stemmed from the inability to discover places within Germany that were using a disproportionate amount of electric power:

> The evidence at hand seems to indicate, in spite of some pessimistic statements put out by German government officials, presumably with the hope of creating the impression among her enemies that Germany is suffering a shortage of electric power, that the fact is that she has an ample power supply.
>
> It is known that many of the power plants in Germany and in occupied territory have been increased in capacity and a considerable block of power is being imported from Belgium and possibly from other neighboring countries. All of the power stations which have been increased are at strategic locations for fuel supply or hydro-power. None of the known additions are of sufficient capacity to point to a development of the kind we are anxious to locate.[13]

What this meant for Bush and the Americans was that there was no way to use the Germans' power supply as a means to discover the location of German atomic weapons research. A power plant dedicated to providing power for atomic research could be supplied from the existing power grid, and therefore the Americans, according to Bush, could not "expect to discover the location of such plants by looking for corresponding large additions to nearby electric power stations."[14]

As the months elapsed, and as American scientists became more and more desperate to discover any tangible information about the German atomic bomb program, suggestions for plans of action became more ambitious. By the autumn and winter of 1942, American scientists had shifted their proposals from an emphasis on traditional intelligence analysis to one that can only be described as an embryonic form of scientific special operations. This began as relatively modest proposals, such as sending scientists to neutral Switzerland to collect German and French scientific periodicals. It was suggested that American scientists overseas could also contact European scientists who might have information on German progress. If the United States could send a qualified scientist who had strong ties to European science and European scientists, some real progress could be made in learning the secrets of the German bomb.

However, it was not to be. The support for these uncomplicated missions (which had at least a moderate chance of success) would later evolve into an elaborate and ruthless proposal in December 1942 to kidnap the German physicist Werner Heisenberg. The idea was first broached in late October, when the Austrian-born American physicist Victor Weisskopf and the German-born American physicist Hans Bethe learned that Heisenberg would be giving a lecture in Zurich, Switzerland, that December. (The news came from the Princeton physicist Wolfgang Pauli, who heard it from the German physicist Gregor Wentzel in Zurich, who learned it from two visiting physicists, a man named Wefelmayer and the Italian physicist Gian Carlo Wick, both students of Heisenberg. Thus went the scientific underground.)[15] Initially, Bethe and Weisskopf discussed only the idea of sending someone to talk to Heisenberg, to learn his commitment to the building of the Nazi bomb. This idea was quickly dismissed as it immediately became apparent that it had a major drawback. Any attempt to talk to Heisenberg about the atomic bomb would reveal what was perhaps the most highly guarded secret of the war: that the Allies were aggressively pursuing an atomic bomb of their own. If Heisenberg was a true believer in the Nazi cause, speaking to him would gather very little intelligence while giving away the crown jewel of U.S. wartime scientific research. Kidnapping Heisenberg was the only prudent action, but Bethe and Weisskopf were not intelligence operatives. They had no experience in planning such a risky operation, and the plan they developed, in which Weisskopf himself would travel to Zurich and kidnap Heisenberg, was impractical at best, and more likely doomed to failure from the very start. Yet Bethe and Weisskopf were so afraid of the progress of German atomic science that they passed this idea up the chain of command, through Oppenheimer and Bush to the military authorities, who saw it for the foolhardy plan it was and rejected it outright.[16] Years later, Weisskopf would write that he felt fortunate that "this ill-conceived plan never took place." He also would wonder how he "could have proposed such a harebrained idea, let alone considered participating in its execution."[17]

By the end of 1942, it had become apparent to most American scientists that little, if any, verifiable or actionable intelligence had been obtained by the United States and its scientific leadership. In a report dated December 15, 1942, Bush, in his role as chair of the Military Policy Committee on Atomic Fission Bombs, informed Vice President Henry Wallace, Secretary of War Henry Stimson, and army chief of staff Gen. George Marshall that despite their best efforts the U.S. scientific community did "not know, unfortunately, just how much progress [the Germans] have made." Bush explained that the subject of German atomic research "is an exceedingly difficult one on which to obtain information as to enemy activity," and acknowledged with resignation, "It must be realized . . . that almost no real information is available," and that any attempt to estimate when a German bomb would be available would be purely speculative. Bush's report did include a guess, however, and it was a number that would have been familiar to anyone who had been involved in approximating German progress in June. "It is entirely possible . . . that [Germany] may be six months or a year ahead in the over-all program due to the head start," Bush said.[18]

Six months of intelligence work had accomplished little, and this unfortunate trend would continue into the first half of 1943. The problem was that American scientists were amateurs when it came to intelligence work; they did not possess the qualifications or experience to accomplish such a difficult task. In their defense, the American scientific leadership was well aware of their shortcomings. Early on in the process of developing the scientific intelligence apparatus, National Defense Research Committee chair James Conant's assistant Harry Wensel suggested that they ask for help: "Should anything be done," he asked Conant, "in regard to bringing in professionals to advise on what to do and how to do it?"[19] The "professionals" Wensel was referring to were the already established members of the American intelligence community: the U.S. Army's intelligence branch (G-2), the Office of Naval Intelligence (ONI), and the Office of Strategic Services (OSS). While his question seems logically sound, particularly considering that the principal problem that scientists were having was due to their lack of experience and knowledge in all things intelligence related (such as analysis and operations), G-2, the ONI, and the OSS actually had little to offer.

The established intelligence apparatus was perhaps less qualified than American scientists to accomplish the task of discovering the progress of the German atomic bomb program. There were three significant reasons for this. First, these agencies were understaffed and overburdened as it was, without adding an atomic intelligence mission to their purview. The second reason was, and still is, a common impediment to effective scientific intelligence: just as scientists are usually not also capable intelligence operatives, so too do most intelligence professionals lack the adequate scientific knowledge to effectively collect and analyze scientific intelligence. The average intelligence officer was unlikely to recognize the importance of key scientific information. Furthermore, educating intelligence operatives as to what to look for scientifically, assuming it could be done sufficiently in a short time, greatly increased the likelihood that information about the United States' interest in atomic weapons would be leaked to the Germans. The fewer people who knew of the U.S. program, the better the chance it stayed secret.[20]

Bush and Samuel Goudsmit would highlight this sentiment in their memoirs written after the war. According to Bush, "Scientific intelligence is not conducted well by Mata Hari methods or through agents who know no science, and there is just as much danger of placing scientific intelligence in the hands of those who do not understand as there is in placing any other part of science in the same tender care."[21] Goudsmit was even less equivocal in his indictment of unqualified intelligence operatives conducting scientific intelligence: "Ordinary Intelligence information yielded nothing of value. There were always fantastic rumors floating around about terrifying secret weapons and atom bombs which were duly reported by the O.S.S. and British agents, but invariably the technical details were hopelessly nonsensical. The reason was obvious. No ordinary spy could get us the information we wanted for the simple reason that he lacked the scientific training to know what was essential. Only scientifically qualified personnel could get us that and a Mata Hari with a Ph.D. in physics is rare, even in detective fiction."[22]

The third problem with the established intelligence organizations was bureaucratic and institutional. The U.S. intelligence agencies carried out atomic intelligence as part of their general activities, and there was no concerted effort to focus particular attention on the German atomic bomb program. Instead, G-2, the ONI, and the OSS (along with several other smaller intelligence agencies within governmental organizations such as the State Department) gathered what Leslie Groves called "scraps and bits of information within the enemy nations that might be useful in adding to the atomic picture."[23] In addition, there was no coordination of effort among the agencies, and no unified command. G-2 was under the direction of the Department of War, while the ONI was controlled by the Department of the Navy and the OSS was under the auspices of the Joint Chiefs of Staff. In theory, these agencies would have been coordinated under the greater war effort against the Axis powers, but in practice there were significant gaps in intelligence coverage. These were further exacerbated by bureaucratic infighting among the organizations—"frictions" (in Groves's euphemistic term) that were particularly acute between the established intelligence organizations of the ONI and G-2, and the newly formed OSS. Trust and cooperation were strained at the highest levels and almost nonexistent at the operational level.[24] These frictions meant that the existing intelligence agencies could not coordinate efforts to discover the extent to which German atomic research had progressed.[25]

Thus, by the summer of 1943 the U.S. military, political, and scientific leadership found themselves in the unenviable position of those who had spent a year fixated on the possibilities of a German atomic bomb, yet had little to show for it. Their understanding of the scientific situation in Germany was roughly the same as it had been in June 1942: the Germans wanted to build an atomic bomb and they had competent scientists, state-of-the-art facilities, all the necessary materials, and the backing of the German High Command. Besides this, however, U.S. intelligence (whether conducted by professionals or amateurs/scientists) could provide almost nothing in the form of actionable information. Something clearly needed to be done to bridge the chasm between those trained in the sciences (those who knew what to look for) and those trained in the craft of intelligence (those who knew how to look). The ideal solution would be for a single individual or small group of individuals to amalgamate the scientific and intelligence fields into one concentrated scientific intelligence organization. This individual or group would be well versed in the scientific fields that atomic theory encompassed, and would also have a general knowledge of the intricacies of intelligence operations. Fortunately the United States had such an individual, and he was close at hand: Brig. Gen. Leslie Groves.

Groves Takes Charge

In September of 1943, Gen. George Marshall, the army chief of staff, asked Groves "whether there was any reason why [Groves] could not take over all foreign intelligence" in the atomic field.[26] After Groves agreed, he and Marshall notified the leadership of the ONI, G-2, and the OSS that Groves would be heading all atomic intelligence from that point on, and that they should give him their full cooperation.[27] The weight of Marshall's office and the respect he had garnered while army chief was enough to mitigate the potential problems that Groves (a newly promoted one-star general) could have faced from the higher-ranked officers who led the various intelligence agencies (both G-2 and the ONI were commanded by two-star major generals, while the OSS was commanded by William Donovan, also a general, but more importantly the most-decorated American soldier in the First World War and someone handpicked by President Roosevelt for the job). For the duration of the war, Groves counted on, and received, the full assistance and cooperation of each of the intelligence chiefs to the utmost of their abilities.

Groves, however, could not solve the most pressing problem: the lack of any significant information about the German atomic bomb program. Groves had followed the progress of the previous year's efforts to learn about German atomic development, and he was well aware of their futility. The scientists, for all their limitations, had done all that could be done with the available intelligence that could be collected through passive methods (collecting information that is delivered, through whatever means, rather than actively sought, mainly through agents in the field and clandestine operations). He believed that the only way to be sure about the progress of the German program was to send intelligence operatives to Europe to learn firsthand, from the European scientists themselves, how far the Germans had gone (this idea would materialize later as the Alsos Mission—see chapters 3 and 4).[28] Yet Groves understood that before this could occur, five major organizational and infrastructural challenges had to be overcome: (1) the full and complete consolidation of authority for atomic intelligence operations and analysis under his immediate command; (2) the creation of an efficient and streamlined command-and-control apparatus for the European mission that would include a centralized information clearinghouse to facilitate the effective targeting of German scientists, facilities, industrial centers, and fissile materials; (3) the exploitation of British scientific and atomic intelligence to its full extent; (4) direct military and covert action that could slow the progress of the German atomic bomb program, allowing time both for the intelligence system to come up to task and for the U.S. bomb program to come to completion; and (5) the prevention, as much as was possible, of any information about the Manhattan Project or the United States' interest in the German program leaking to anyone outside of the need-to-know. Once these tasks were initiated, Groves could begin putting in place the pieces that would make up the Alsos Mission. Until then, there was much work to be done.

Groves possessed a well-defined management philosophy. He believed in the paramount importance of the centralization of authority. As he put it, "If there was one guiding principle throughout [his command of the U.S. atomic bomb project], it was that those who carried responsibility were possessed of corresponding authority."[29] Up to the fall of 1942, Groves was in charge of a number of major construction projects in the United States, most notably the construction of the Pentagon. He was in direct command of a force of one million men in the Army Corps of Engineers, who were building facilities in the United States at the rate of $600 million in construction costs each month. Thus, Groves possessed the experience of leading large projects, managing significant budgets, and managing disparate personalities, traits that were imperative for the successful completion of a U.S. atomic bomb. Groves wrote that when he was selected, he was told "to take complete charge of the entire Manhattan Engineer District project,"[30] instructions that he took to heart.

This process would begin with his appointment in September 1942.[31] When Groves took command, he dramatically accelerated the transition of the responsibilities for atomic research and development from the OSRD to the War Department and the Army Corps of Engineers. In doing so, he converted to his control not only laboratory work, but also pilot plant construction and full-scale production authority. At the same time, a supposedly overarching control organization, the Military Policy Committee, was created by Secretary of War Henry Stimson. Despite the fact that all of the members of the committee outranked Groves,[32] he quickly maneuvered himself into the leadership role. The committee itself was originally intended to have nine members, but Groves persuaded Stimson that a committee of that size would be inefficient and impractical.[33] Committee meetings were always held in Groves's office, ostensibly because his office was, according to Groves, "where needed papers were readily available and security difficulties were minimized."[34] Perhaps this was truly the case, but before long Groves began to dominate these meetings and their agendas. While the committee was intended to be a consortium of especially qualified men who would come to consensus, if not unanimity, on U.S. atomic policy, Groves was not only the most important voice on policy, but before long the sole arbiter of the future of U.S. atomic research and development. According to Groves, "As time went on and I became much more familiar with our operations than any of the others, it became more and more a question of approval and of discussion rather than of decision."[35]

Groves's leadership style could be kindly described as hands-on, or more critically as micromanagement. Delays were not permitted, and any officer who felt that a problem would require the project to be delayed by as much as twenty-four hours was required to immediately report that issue to Groves.[36] His infamous compartmentalization of information, primarily for security purposes, was also used so that Groves could avoid delays by keeping the scientists working only on their own projects, and not those of their peers. This meant that Groves could, he wrote, maintain full control over the subjects his scientists were researching, preventing them from spending "time and effort thinking about the responsibilities of others," which could allow any devolution of attention "into futile sidelines."[37]

Groves also purposely kept his staff small. He did not have a chief of staff or an executive officer, and for the majority of the war his office in Washington, DC, consisted of only five rooms. He did not need more space, since he rarely spent his time "sitting in Washington reading committee reports, holding press conferences, or making speeches."[38] He was in constant contact with all Manhattan District operations, either in person or by telephone. Groves's theory on staff was simple, and encapsulated his management philosophy: "As soon as the staff gets too large, it begins to operate independently and trouble is certain."[39] As a result, Groves was reluctant to delegate authority to others, and only did so when he had complete trust in both the competency and the loyalty of that individual. The closest he came to having a second-in-command was the district engineer Col. K. D. Nichols, who mainly served as an administrative aide but who would maintain local operational control over a Manhattan Project site such as Oak Ridge, Tennessee, or Hanford, Washington, when Groves was working elsewhere. While the relationship between Groves and Nichols was never officially defined in writing, Groves later insisted that while Nichols did have wider latitude than most others to act without direct orders from him, Nichols was always under his "complete control."[40]

Gradually throughout the war, Groves would expand his responsibility and authority to include project security, counterintelligence, the selection of Japanese cities for targeting, the arrangement of logistical support from the military for bases of operation overseas, and, of course, intelligence on atomic research and development in Germany.[41] Immediately upon accepting the job as atomic intelligence chief, Groves sought and received assurances from the army and navy intelligence branches that they would redirect any atomic intelligence they collected to Groves and the MED. He then personally visited the director of the OSS, Maj. Gen. William Donovan, to coordinate the activities of the OSS with the objectives of U.S. atomic intelligence. At the meeting, which also included Donovan's executive officer, Col. G. E. Buxton, Donovan assigned one of his top officers, Lt. Col. Howard Dix, to act as liaison with the MED and, Groves wrote, "to ensure that all atomic information collected by OSS would be forwarded promptly to the Intelligence Section [of the MED]." In his effort to establish a personal relationship with the OSS, Groves's hands-on approach truly paid off. According to Donovan, Groves was the first general officer to have ever met with him in his office. On all other occasions, a general had sent an aide or subordinate to meet with Donovan (a clear indication of lack of respect, one of the "frictions" that existed between the other U.S. intelligence agencies and the OSS). Colonel Buxton would later tell Groves that his personal outreach to the OSS "ensured the utmost in special treatment for the MED."[42]

As much as he might have liked to, Groves could not direct the day-to-day operations of the research and development of the U.S. atomic bomb and simultaneously oversee internal security, counterintelligence, and foreign intelligence. He could count on Colonel Nichols to assist him in commanding the bomb production program. In the same way, he could depend on a trusted subordinate, Lt. Col. John Lansdale Jr., to oversee all MED intelligence operations. According to Lansdale, he "had a very close relationship with General Groves and saw him virtually every day that he and [Lansdale] were both in Washington and [they] not infrequently traveled together."[43] A graduate of the Virginia Military Institute and Harvard Law School, Lansdale was a reserve artillery officer in the 1930s. As it became more and more likely that the United States would soon enter the Second World War, Lansdale decided to request a call to active duty to serve in the Military Intelligence Division of the War Department General Staff (G-2). He reported on June 10, 1941, as a first lieutenant assigned to the Investigation Branch, Counter-Intelligence Group.[44] After distinguishing himself as the G-2 representative on what he described as the "Japanese-American Joint Board responsible for releasing Americans of Japanese extraction on an individual basis from the concentration camps in which they had been confined during the period of hysteria immediately after the Pearl Harbor attack,"[45] Lansdale first came to the attention of Groves when Lansdale was asked by James Conant in February 1942 to investigate the security at the Radiation Laboratory at the University of California, Berkeley. Lansdale's competence, thoroughness, and discretion in performing his duties convinced Groves to put him in charge of all MED security when Groves took responsibility for the project in September 1942.[46] At this point, Lansdale was still technically a member of Army G-2, but this would formally change in the winter of 1943–44 when the MED and Groves were given the task of directing all aspects of atomic intelligence. Groves requested and was granted Lansdale's transfer to the Army Corps of Engineers so that Lansdale could devote his full time to the atomic bomb project.[47]

Lieutenant Colonel Lansdale was directed to take command of each of the three key aspects of the MED's atomic intelligence program: internal security, counterintelligence, and foreign atomic intelligence. Serving under both Groves and Lansdale was Maj. Robert Furman, the head of the MED's foreign intelligence operations. Furman, a graduate of Princeton University and a civil engineer by training, was a veteran of the Groves command hierarchy, having served under him during the construction of the Pentagon. When Groves was appointed to take control of the MED, he brought Furman, one of his closest and most trusted aides, with him. More than anyone else, Furman was responsible for directing the day-to-day atomic espionage operations carried out against the Germans.[48]

Another important task for Groves was to establish an efficient organizational foundation for the mission to Europe. Primarily this would be the creation of a command infrastructure, but it would also include building a framework of intelligence information on the German program in order to more effectively initiate, and then to sustain, operations in Italy, France, and eventually Germany. Even with intelligence professionals who were well versed in the intricacies of atomic research, however, this was no easy task. The Gestapo, the German counterespionage service, was able to block access to secret scientific and technical information. According to the historian John Keegan, "There was little that was romantic about spying in Hitler's Europe. The business was furtive, nail-biting and burdened by the suspicion of betrayal. Many agents were betrayed." Despite this, considerable information about the German rocket program (the V-1 and V-2) did manage to make its way to the Allies. Primarily this was due to the fact that these programs required test flights that could be observed by those willing to transmit that information to the British.[49]

Ironically, it was this wealth of information about German rockets that would feed the paranoia about the German atomic bomb program. Since very little intelligence about the German uranium project was collected by the United States, the perception was that the Germans felt the program to be so important to the war effort that they were placing a concerted emphasis on project secrecy. The Americans also assumed that when information trickled out of Europe suggesting that the German program was not as advanced as was feared, this was a deliberate disinformation campaign to mislead the Allies. In particular, when American scientists read German scientific publications that were smuggled out of Europe through neutral countries, they noticed that they contained a number of articles written by prominent German scientists on topics that in the United States were forbidden by governmental censorship (for instance, articles on atomic physics or isotope separation). Instead of accepting this for what it was—an indication of German lack of interest in atomic weapons—the U.S. scientific intelligence community concluded that the only justification for the Nazi regime to allow the publication of such key scientific information was to trick the Allies into believing that Germany had given up on its atomic ambitions.[50]

Successfully differentiating between actionable intelligence, disinformation, and random, inconsequential data was (and still is) a major challenge for even the most capable and experienced intelligence analyst. Groves, Lansdale, and Furman had to meet this challenge with the added obstacle of operating in an intelligence environment (atomic physics) that was constantly changing and, because it was a newly developed and evolving science, ill-defined. To their credit, they were cognizant of a common intelligence pitfall that could have derailed the fledgling atomic intelligence program at an early stage: assuming your enemy will act in the exact manner you will. In intelligence terminology this is called "mirror imaging," and in this case, Groves feared that the Germans might develop methods leading to a bomb that were dramatically different from those the American scientists had devised. Groves and the others understood the theoretical, technical, and production difficulties American scientists and engineers had been facing, yet they recognized that, as Groves put it, the "chief danger was that [the Germans] might come up with relatively simple solutions to the problems we were finding so difficult."[51] Robert Oppenheimer told Furman that despite the Americans' assertion that massive separation plants were necessary to amass significant quantities of U-235, the Germans could potentially have found a way to do it more cheaply and efficiently. Isotope separation was a new science, and therefore a German scientist might, according to Oppenheimer, "come up with a way to do it in his kitchen sink."[52] Groves worried that the Germans would also discover a faster and better way to produce plutonium, in particular because they had a distinct advantage not shared by their American and British counterparts:

> It had always seemed to most of us that their best prospects lay in the use of plutonium, which would demand a much smaller industrial effort as well as considerably less in the way of time, critical equipment and materials than any other method—provided they were willing to ignore safety precautions. This I felt the Germans would do, for considering what we already knew of their treatment of their Jewish minority, we could only assume they would not hesitate to expose these same citizens to excessive radiation. Hitler and his ardent supporters, we felt, would consider this a proper use for an "inferior" group, quite apart from the saving in effort and materials and time.[53]

Ultimately, it was up to Furman (under the guidance of Groves and Lansdale, of course) to determine what information to ignore, what information to pay heed to, and what avenue of approach should be taken in regard to U.S. scientific intelligence operations. To fully prepare himself for this task, Furman met with several American atomic physicists to gather their advice on how to proceed. They advised him to pay close attention to German scientific journals not only in atomic physics, but also in other fields of science, such as electronics and chemical engineering, that would be integrated into any German bomb program. In addition, Furman was told to follow the development of German industry, construction, and mining in the fields relevant to atomic research, and to try to ascertain connections between scientific, engineering, and industrial personnel,[54] which would be, according to the author Jeffrey Richelson, "a sign that an attempt was being made to transform theory into an atomic capability."[55]

Most notably, Furman met with Robert Oppenheimer, who provided him with a watch list of Germany's leading scientists in the fields of atomic physics and chemistry. In a follow-up letter to Furman, Oppenheimer reinforced his proposition that the key to learning the status of the German program was its scientists. He emphasized the necessity of discovering "the whereabouts and activities of the men who are regarded as specialists in this field and without whom it would certainly be difficult to carry out a program effectively." He also argued that German raw materials acquisition and plant construction could be avenues of discovery. Uranium was the obvious material to keep track of, but according to Oppenheimer, U.S. surveillance of German raw materials acquisition should not be limited to uranium; the United States should also pay attention to German interest in graphite, beryllium, and heavy water (each of which could be used as a moderator).

Also, U.S. intelligence should look for large plant construction that would require conspicuous amounts of power. Perhaps a major chemical company like I. G. Farben, Oppenheimer wrote in a letter in September 1943, could have been contracted to build such a plant. If so, this would present a serious obstacle to discovery, since "it would be quite possible to conceal the plant among other war projects." A possible recourse, he said, could be "to investigate the radioactivity of rivers some miles below any suspicious and secret plant." Scientists back in the United States could then determine whether this was the home of a German atomic reactor.[56] Of course, this method would involve agents on the ground inside Germany, not yet a reality at the time of Oppenheimer's letter.

Centralization and Integration

Yet Furman knew the mission was to begin soon enough, and he had more work to do before it could begin with any chance of success. A key element of any operation of this magnitude would be to maximize the contributions (as much as they could be within security restraints) of the entire U.S. government. Thus, armed with a letter from the commander of G-2, Major General Strong, Furman contacted all of the other governmental agencies that might be able to help him determine the extent of the German atomic bomb program. He told these agencies that he was interested in information on earthquakes (or other significant seismic activity), large industrial facilities, movements of scientists, and industrial plants with little in the way of visible production (a bomb factory would have a significant amount of raw materials going in, but nothing of substance—except, in the end, an atomic bomb—regularly coming out).[57] Furman also met with OSS chief Donovan in October 1943. Up to this point, the OSS had been passively collecting intelligence about the German bomb program while carrying out its normal clandestine activities. Furman, through the guidance of Groves, pressed Donovan to begin an active campaign. Donovan agreed, and created a section of the OSS Secret Intelligence branch to pursue information on German scientists, industrial plants, and research. In November, Donovan instructed Allen Dulles, the OSS chief of station in Bern, Switzerland, to seek out information on a number of Italian scientists, including Gian Carlo Wick and Edoardo Amaldi (see chapter 1).[58] Throughout the remainder of the war, the OSS would play an invaluable role in the scientific intelligence collection effort.

Groves, Lansdale, and Furman were doing all they could to establish an effective intelligence organization in Washington, but they understood that eventual success would require a presence closer to the theater of operations. In January 1944, Groves sent Maj. Horace K. "Tony" Calvert to London to establish a forward base of operations for the MED's intelligence effort. Calvert, a lawyer in the oil industry in peacetime, was chosen for this assignment because he had significant experience in intelligence (he had worked under Lansdale when Lansdale was at G-2, and then followed him to the MED), and had an extensive background in the U.S. atomic bomb project. As a result, Groves felt "that he would be well qualified to recognize any danger spots in the German picture." Groves gave him the following instructions as he departed for Europe:

> He was to gather all possible information on the various atomic energy efforts under way in Europe, particularly those being carried out by the Germans; to make use as far as possible of existing American and British channels; to keep his intelligence estimate up to date at all times and to report to us in Washington everything that he considered to be of importance. He was also expected to establish close and friendly relations with the Englishmen and Americans with whom we might have to deal from time to time, both in London and, as the situation developed, on the Continent.[59]

Calvert arrived in London and immediately reported to Col. George B. Conrad, G-2 of the European Theater of Operations, U.S. Army. Armed with a letter of introduction from General Strong, Calvert was able to convey the importance of his mission and, as a result, was given a desk in Conrad's office where he could organize and analyze the raw intelligence data as it arrived. Afterward, Calvert reported to John Gilbert Winant, U.S. ambassador to the United Kingdom. Unlike Conrad, Winant was only given incomplete and misleading information about Calvert's mission (for security reasons, atomic information was considered need-to-know, and Winant did not need to know), but it was enough to get Calvert the promise of the utmost support from the ambassador, a desk in the embassy, and the official title of assistant military attaché. Soon thereafter, Calvert was joined in the embassy by another MED intelligence officer, George C. Davis, and accompanying support personnel (three Women's Army Corps members—WACs—and two counterintelligence agents).[60]

Once the organizational details had been settled, Calvert and his team began the arduous task of analyzing all the known intelligence on the German program. To do so, they combined what had been collected and extrapolated by various methods in the United States with what they could themselves collect from German refugees who had immigrated to London. They also studied German physics journals, interviewed anti-Nazi scientists in neutral countries, and perused German newspapers to obtain clues as to the locations of top German atomic scientists. Calvert's team knew from Allied intelligence that many of Germany's leading scientists were working on the secret rocket program in the research institute at Peenemünde, but they believed that no nuclear scientists were among them. Therefore, the majority of prominent German atomic scientists—Heisenberg, Hahn, Harteck, Weizsäcker, et al.—remained unaccounted for. Operating on the principle that if he could find some of them, those would then lead to the rest (since it was more than likely that a program of the size required to build an atomic bomb would be a collaborative effort), Calvert focused his attention on locating the very top echelon of German atomic physics. In time, this process proved successful; U.S. intelligence personnel, according to Groves, were able to acquire "recent addresses for a majority of the scientists in whom [they] were interested."[61]

Scientists were only one aspect of the overall picture. Calvert and his team also studied German raw materials acquisition and industrial production. They analyzed uranium supplies at mines and processing centers controlled by the Germans, particularly the mine at Joachimsthal, Czechoslovakia, and a prominent uranium processing center outside Berlin. By utilizing various methods of observation and analysis, such as studying aerial surveillance photographs for activity at the mines, microscopically measuring the amount of ore piled outside the mines on subsequent days, and knowing the general grade of the ore extracted, the team could extrapolate the mine's rate of production. They then could study laboratories and industrial plants in much the same way. According to Groves, "Lists were compiled of all of the precious metal refineries, the physics laboratories, the handlers of uranium and thorium, manufacturers of centrifugal and reciprocating pumps, power plants and other such installations as were known to exist in the Axis countries." The team systematically checked each plant on the list, only eliminating it when it had been proven that it was not being used for atomic research and production. Any facility that remained on the list was thoroughly vetted in any number of ways, including aerial surveillance, by the OSS and other intelligence agencies, and by the various underground or partisan movements. Whatever remained after this extensive process would be a future target of the scientific intelligence mission. "By hard work and constant effort," Groves wrote, "Calvert was ready by the time [the scientific intelligence mission to France] reached Europe on the heels of the invading armies with a good list of the first intelligence targets, dossiers on all the top German scientists, where they worked and where they lived, the location of the laboratories, workshops and storage points of interest."[62]

While Calvert was establishing the MED's scientific intelligence operation in London, Furman was in Washington exploiting a new and significant source of intelligence. Niels Bohr, Nobel laureate and close confidant of Heisenberg's, had immigrated to the United States through Great Britain in December 1943. Bohr gave Furman information on a number of German scientists, including their activities, associations, and political persuasions (that is, how they felt about the Nazis). Most notably, however, Bohr related to Furman the details of a meeting between Bohr and Heisenberg that took place in Copenhagen in September 1941. What was actually said at this meeting is the source of some historical controversy. According to historical accounts from Robert Jungk, David Irving, Thomas Powers, and Heisenberg's wife, Elisabeth,[63] Werner Heisenberg met with Bohr in order to deescalate tensions between German and Allied scientists, and ultimately to ensure that atomic weapons were not used in the war by either side. Each account portrays Heisenberg as a reluctant participant in the German bomb program and insinuates that he, along with many other prominent German physicists and chemists, was doing what he could within the restraints of the Nazi system to retard the progress of atomic research.

Jungk's 1956 book, *Brighter Than a Thousand Suns: A Personal History of the Atomic Scientists* (published in German and first translated into English in 1958), which Heisenberg himself collaborated on and contributed to, argues that "although Heisenberg may not have longed for the eventual German collapse, he was convinced, purely as a matter of logic, that Germany must lose."[64] Jungk also contends that Germany had given up on building an atomic bomb as early as late summer 1941 (before the Heisenberg visit to Copenhagen), and that Heisenberg, known to give public statements in defense of the Nazi regime, only did so "in order to disguise his true sentiments" from the German authorities.[65] In a letter from Heisenberg to Jungk included in the book, Heisenberg wrote that by the time he met with Bohr, German scientists were working to ensure that Germany would never have an atomic bomb. The difficulties and immense resources involved in building a bomb, he said, "enabled the physicists to influence further developments. If it were impossible to produce atomic bombs this problem would not have arisen, but if they were easily produced the physicists would have been unable to prevent their manufacture. This situation gave the physicists at that time decisive influence on further developments, since they could argue with the government that atomic bombs would probably not be available during the course of the war."[66]

Heisenberg's recollections are based on extensive notes he wrote after the meeting, and according to Jungk "are the best existing source" for what happened in Copenhagen in the autumn of 1941.[67] Yet what Jungk did not know was that both Niels Bohr and his son Aage (who would also become a Nobel laureate in physics) had recorded a much different account of the events in Copenhagen, and it can be safely assumed that it was Bohr's version of the story that was related to Furman in December 1943. Aage Bohr contends that Jungk and Heisenberg's account of the meeting "has no basis in the actual events," and that instead of emphasizing that the Germans had quit their atomic ambitions, Heisenberg gave his father "the impression that the German authorities attributed great military importance to atomic energy."[68] In an unsent letter to Heisenberg written after the war, Niels Bohr wrote that he was "greatly amazed to see how much your memory has deceived you in your letter to [Jungk]." He also revealed that during his visit Heisenberg conveyed his "definite conviction that Germany would win and that it was therefore quite foolish for [Bohr and the Allies] to maintain the hope of a different outcome of the war." In addition, he told Heisenberg, "You spoke in a manner that could only give me the firm impression that, under your leadership, everything was being done in Germany to develop atomic weapons and that you said that there was no need to talk about details since you were completely familiar with them and had spent the past two years working more or less exclusively on such preparations."[69] Most damningly, Bohr insinuated that Heisenberg was sent to Copenhagen by the German authorities, and that the meeting in 1941 was "boldly arranged" to either discover what Bohr knew about the Allied bomb program or convince the Allies to give up their ambitions altogether.[70]

For present purposes, it does not matter what actually occurred in Copenhagen. What does is what Bohr told Furman in December 1943. If it was consistent with what he and his son argued after the war, then Furman and U.S. scientific intelligence were given valuable insight into the status of the German atomic bomb program, and its leader, Werner Heisenberg.

Part of creating the overall intelligence picture of German atomic research included exploiting resources within British intelligence. One of John Lansdale's first priorities after permanently joining Groves and the MED at the end of 1943 was to establish a relationship with his British counterparts. In January 1944 Lansdale sent Majors Furman and Calvert to London to make contact with British Secret Intelligence and those scientists and technicians working on the British version of the MED, a program the British code-named "Tube Alloys." There they met with Sir Charles Hambro, a member of a prominent British banking family and an experienced intelligence officer.[71] Furman and Calvert also established a relationship with Michael Perrin, the administrative head of the Tube Alloys office; his assistant, David Gattiker; and Lt. Cdr. Eric Welsh, a British Intelligence officer specializing in foreign atomic development, particularly in Norway (he had served as head of the Norwegian Section for British Intelligence). Furman would soon return to Washington, but Calvert remained and acquired a desk (his third in London) in the British Atomic Energy Office, where he acted as the liaison between Groves and the British.[72]

The British scientific leadership had convinced their government to take the German threat seriously more than two years before the U.S. scientific intelligence program would begin. In April 1940, the Military Application of Uranium Detonation (MAUD) Committee, made up of prominent British scientists with close ties to the government, met for the first time and agreed to establish an aggressive program to research atomic development.[73] They also decided almost immediately to start a program to monitor German development in atomic physics, identifying a list of German scientists who would be instrumental to any Nazi atomic bomb, including Heisenberg.[74] They followed that with a study of German scientific periodicals, obtained through neutral countries, and a systematic search of course offerings in German universities for classes taught by key atomic physicists.[75]

In addition, the British had followed German procurement of raw materials for bomb development. They discovered that the Germans had captured the largest reserve of uranium oxide in Europe when they occupied Belgium in 1940. This material was located at the refinery of the Belgian company Union Miniere, in Oolen, a small town northwest of Brussels. The British also learned that the Germans were trying to increase the production of heavy water at the Norsk Hydro facility at Rjukan in occupied Norway. Taken together, these indicated that the Germans were well equipped to undertake an aggressive bomb project. Yet the British had no indication of a link between the scientists, the raw materials, and the large-scale industrial effort necessary to build an atomic bomb. Finding such a connection would occupy their efforts for the remainder of the war.[76]

Major Calvert's mission as it regarded the British was to tap into this established intelligence infrastructure and send back to Washington any information the British gathered about the German bomb. Nominally, the relationship between the U.S. and British scientific intelligence operations was intended to be a joint partnership on equal footing. However, both sides assumed they would take the leading role. The British believed their significant experience in scientific intelligence would convince the Americans to let them maintain their dominant position. The British physicist R. V. Jones writes that after they first met with Furman and Calvert and realized the Americans were novices in scientific intelligence, the British celebrated "in anticipation that [they] were so obviously going to be the senior partners in the exchange." Yet Groves was not about to cede power to anyone, certainly not the British. He saw this relationship not as a two-way exchange of information, but instead as an opportunity for U.S. intelligence to expand its sources of intelligence collection. The United States would take what it could from British Intelligence, and only reluctantly send back such information as it must to ensure a continuing relationship. For example, the planning and preparations for the Alsos Mission were completed independently of British involvement. The British were informed after the fact when the scientific chief of the Alsos Mission to France, Samuel Goudsmit, arrived in London and explained his mission to British scientists and intelligence officials. In a fleeting moment of resentment toward their American allies, the British considered forming their own scientific intelligence field team, independent of and as a competitor to Alsos. Yet the reality of the situation would soon sink in, and the British decided, according to Jones, that "it would be best for Anglo-American relations if, despite our greater experience, we should seek American *permission* to join the ALSOS mission under American leadership, and thus become very much the junior partner" (emphasis added).[77]

The British, somewhat counterintuitively (particularly considering their vulnerability to German air attack), were far less concerned about a German atomic bomb than were Groves and the MED intelligence team. There were two reasons for this. First, the British codebreakers at Bletchley Park had deciphered tens of thousands of German messages sent over Enigma machines, and paid close attention for any mention of a German atomic bomb program, uranium, Heisenberg, or any kind of large or special industrial development program. In the past, they had deciphered messages relating to the German rocket program and other equally secret enemy projects, yet nothing had come across Enigma that even remotely referred to German atomic ambitions. The British concluded that this absence of communication indicated that the Germans were not actively engaged in a large program to develop atomic weapons. Second, while the British shared their *opinion* on this issue with the Americans, they did not share the *source* of their information. Calvert, Furman, and Groves were given the take (or the collections product) from Ultra (the code name given to the signals intelligence collected through the breaking of the German secret codes), but were never told where—or from whom—the intelligence was collected. According to Robert Furman, the perception among the Americans was that the British had not collected significant intelligence on the German atomic bomb program. Instead, the Americans thought they were basing their conclusions mainly on information they had obtained from scientists who had recently escaped occupied Europe, such as Niels Bohr and Lise Meitner. For Groves and his MED intelligence team, this lack of direct evidence—and the apparently dubious source of British intelligence—meant that there was no choice but to continue the planning for the Alsos Mission.[78]

When Groves assumed responsibility for the U.S. atomic intelligence program, he knew that the consensus among the U.S. scientific leadership was that the Germans had a significant lead over the Allies in the development of the atomic bomb. He also understood that it would take some time for his intelligence operation to reach its full potential. Thus it was imperative for Groves to find a way to slow German progress to allow either for the U.S. bomb project to overtake that of Germany, or for U.S. scientific intelligence to grow into an effective and efficient organization (or, in an ideal world, both).

Groves decided that the best means to accomplish this task was through direct military action, specifically a dedicated bombing campaign. He felt that explicitly targeting German-controlled raw materials manufacture, research facilities, and prominent scientists could give the Americans the time they needed to close the atomic gap. The primary target for bombing was the Norsk Hydro heavy water plant located about seventy-five miles west of Oslo, Norway. Producing an estimated 120 kilograms of heavy water each month for the German atomic bomb program, the Rjukan facility was a key component of the German quest for atomic weapons. When Groves had taken control of the Manhattan Project in late summer 1942, he had pushed for British covert action against the plant. The British complied, and in October 1942 they sent four Norwegian expatriate commandos into the Rjukan area to prepare for a larger follow-on force that was sent into Norway on the night of November 19. The mission, however, ended in disaster when the gliders carrying the main force crashed in Norway, killing most of the commandos. Those who survived were quickly captured by the Germans and summarily executed.

British Intelligence, realizing the importance of the facility, decided to send a second mission in February 1943. This time they parachuted a much smaller force, six Norwegian commandos, onto a frozen lake thirty miles north of the plant. Armed with plastic explosives, the commandos attacked the plant on the night of February 27 by sneaking into the plant via a cable intake that led directly to the heavy water containers. The commandos were able to destroy all eighteen stainless-steel electrolysis cells of the high concentration plant, destroying half a ton of heavy water without any Norwegian, or even any German, casualties.[79]

British Intelligence's initial reports estimated that the plant would be out of operation for at least two years.[80] A day later it amended this statement, explaining that its earlier estimate that the plant should be considered "ineffective for at least two years" should instead say that it would "not be fully effective for more than 12 months."[81] In reality, both estimates were wrong, and the plant was fully repaired by April 1943. By the fall of 1943, the plant was gearing up to resume operations at a level commensurate with the production it had achieved prior to the commando raid.[82] German scientists had shipped heavy water from laboratories in Germany to refill the cells and jump-start the rebuilding process.[83] Frustrated by the lack of success of covert action, and buoyed by his new power as commander of U.S. atomic intelligence, Groves demanded direct military action against the plant. In a letter to George Marshall, Major General Strong of Army G-2 explained the rationale for the bombing: "Dr. Bush and General Groves consider it of highest importance that the heavy water plant with adjoining power plant and penstock at Rjukan near Vemork, Norway, which have been restored to operation, be totally destroyed. The destruction of the power facilities as well as the actual manufacturing facilities is desired as this is the only immediately available source of power (DC) necessary for producing heavy water in any quantity. I concur."[84]

General Marshall approved the request, and on November 16, 1943, B-17 bombers from the U.S. Eighth Air Force set off to take Norsk Hydro out of the war permanently. The mission's planners scheduled the attack to coincide with the plant's lunch period, so as to limit the number of Norwegian civilian casualties. One hundred forty B-17s dropped over 350,000 pounds of bombs on the target area, destroying the power station and fatally damaging the electrolysis unit that provided hydrogen to the high concentration plant. While the bombing did not completely annihilate the heavy water plant (in fact, it left much of it untouched), the attack convinced the German High Command to decommission the plant and move heavy water production to a safer location inside Germany. To do so, the Germans planned to dismantle the plant and transport its component parts, and whatever heavy water remained, to a secure location hundreds of miles to the east. The plant components and its valuable heavy water would be vulnerable until they reached the safety of German soil, and this gave Allied intelligence a unique opportunity to significantly reduce the threat of German heavy water.[85]

British Intelligence learned through the Norwegian resistance that the Germans were shipping the heavy water in barrels by rail back to Germany. To do so, they had to first transport the railcars by ferry across Lake Tinnsjø, one of the largest lakes in Norway. A Norwegian commando was able to sneak aboard the ferry, the *SF Hydro*, prior to its departure and plant plastic explosives along its hull. On February 20, 1944, the explosives detonated and the *Hydro* sank, sending German freight cars and thirty-nine barrels of heavy water to the bottom of the lake. Twenty-six of the fifty-three passengers and crew drowned (not counting the Germans escorting the railcars, all civilians), but in the minds of Groves and Allied intelligence this was acceptable collateral damage in the battle for atomic supremacy.[86]

In addition to the Norwegian heavy water facility, Groves advocated a bombing campaign against a number of key German scientific and industrial targets. These would include research facilities such as the Kaiser Wilhelm Institute for Physics, the Kaiser Wilhelm Institute for Physical Chemistry and Electrochemistry, and other scientific centers where atomic bomb research was thought to be conducted. Groves's list also included industrial plants where technologies or materials tangential to atomic research, yet still intrinsic to its success, were produced, such as high explosives. Unlike the Rjukan bombing mission, however, the physical destruction of the plants was only a part of the overall goals of the campaign. To be sure, Groves hoped that bombing could damage these facilities, as Strong summarized, "to put them out of commission for a considerable period of time," but another goal, perhaps the primary goal, was to reduce the scientific capabilities of German atomic physics. To put it another, less euphemistic way (and the way it was presented to General Marshall), "The killing of scientific personnel employed therein would be particularly advantageous."[87]

If Groves did not have to worry about negative repercussions, it is likely he would have specifically targeted every prominent German scientist, laboratory, uranium mine, and atomic bomb–related production facility in occupied Europe, yet he did not have that latitude. It was not for lack of authority. Groves at that point could have asked for, and received, permission to attack almost any target he felt necessary to beat the Germans to the bomb. Instead, the primary factor limiting Groves's actions was the fear that by appearing to give too much attention to German atomic development, Groves might tip off the Germans to the United States' own bomb program. If the Germans discovered the United States was attempting to build an atomic bomb, they were certain to redouble their efforts to build their own. In addition, they would also take steps to conceal the German program from observation and espionage. The United States was having a difficult-enough time learning about German atomic development as it was. If the German atomic bomb program went underground, it would make a difficult task even harder, and perhaps impossible. Thus, Groves was forced to find a delicate balance between aggressively pursuing intelligence on German atomic development and maintaining the secrecy of his own program.

Groves and the Americans were given a direct lesson in the need for discretion in the immediate aftermath of the commando raids on Norsk Hydro. In March 1943, a Swedish newspaper published a report on the raid and the German response. The article speculated that the target of the raid was heavy water, and reported that European and Americans scientists were working on the production of a new secret weapon using the heavy water as a means to achieve a massive explosion. By April, the story had moved across the Atlantic, as the *New York Times* published an article about the raid entitled "Nazi 'Heavy Water' Looms as Weapon." Subtitled "Plant Razed by 'Saboteurs' in Norway Viewed as Source of New Atomic Power," the *Times* article reported that heavy water had "hidden atomic power that can be used for the deadly purposes of war as well as the happier pursuits of peace," and that it "apparently has become a source of anxiety for those Allied leaders who plan attacks against enemy targets." Moreover, the article identified the potential uses of heavy water in producing an atomic bomb: "Heavy water or, more correctly, heavy hydrogen water, is believed to provide a means of disintegrating the atom that would thereby release a devastating power."[88]

Groves was able to contain the damage, mainly through the help of Harold Urey, the discoverer of heavy water, who scrambled to tell anyone in the press who would listen that heavy water could not be militarized.[89] Groves convinced the *Times* not to follow up the story by appealing to the newspaper's patriotism, but this experience left Groves with a lasting impression. When direct military force was used to bomb key German atomic targets (whether scientists, research facilities, or industrial plants), Groves made sure the missions were part of a larger bombing campaign in order to conceal the scientific, atomic-related targets. The B-17s that attacked Norsk Hydro, for example, were joined by several other flights of bombers that attacked targets throughout western and northern Europe to mask the primary target.

Keeping the Secret(s)

A major consideration in keeping the Germans in the dark about the U.S. bomb project (and thus the U.S. atomic intelligence program) was, of course, security and counterintelligence, both of which were under the direction of John Lansdale. Lansdale first gained experience in atomic intelligence matters when he was asked in February 1942 to report to James Conant at the National Defense Research Committee. Conant wanted Lansdale to go undercover into the various laboratories conducting atomic research to determine whether their security was adequate. In almost every case, security was almost nonexistent, forcing Conant to take immediate and punitive measures.[90] Lansdale had also built the MED's security and counterintelligence apparatus while he was still assigned to G-2, since the War Department's Counterintelligence Section was responsible for internal security for the first year of the project. With the full approval of Strong and Groves, Lansdale built an intelligence organization at headquarters, at G-2, and also in the office of each Service Command (Corps of Engineers, Quartermaster Corps, Medical Corps, Signal Corps, Chemical Warfare Service, Ordnance Department, Military Police, Finance, Transportation, etc.) and the Western Defense Command. Each of these offices operated entirely outside regular military channels, and Lansdale maintained separate records and chains of command: a liaison officer in the Service Commands reported to Lansdale, and then Lansdale to Groves. Lansdale put it this way: "Within a comparatively short time we had several hundred officers and agents in this nameless adjunct to the Military Intelligence Service. Thus, the MED and Groves were able to utilize all the resources of the Army counterintelligence organization without having to disclose through regular channels the nature of [their] work."[91]

When Groves took control of all U.S. atomic intelligence, he integrated security, counterintelligence, and foreign intelligence under one command. The timing was fortuitous. By the end of 1943 it was becoming increasingly difficult for Lansdale to carry out his duties while remaining assigned to G-2. The organization he had formed for atomic security, he said, had become "so large that it was almost impossible for it to operate outside of regular channels any more," and a reorganization of the War Department made G-2 less efficient and incompatible with Lansdale's system.[92] Therefore, Lansdale was transferred to the MED, along with the detachment of officers he had cultivated in each of the Service Commands and the Western Defense Command. In all, 148 officers and 161 enlisted men followed Lansdale to the MED.[93]

Together, Groves and Lansdale acted quickly to ensure that the Manhattan Engineer District would remain a mystery to the Germans. They were able to designate as "restricted" the airspace over the three tracts of land most important to the project: Oak Ridge, Tennessee; Hanford, Washington; and Santa Fe/Los Alamos, New Mexico. This had the effect of protecting the work at these sites by forbidding flights over the projects.[94] Lansdale was also able to fend off the Justice Department, which had begun an investigation of the DuPont Company. DuPont was a major contributor to the Manhattan Project, and an overt, public investigation ran the risk of exposing its secret government work to the wrong people. Lansdale used the power of his office (in particular the power Groves had acquired for himself) to convince Tom Clark, assistant attorney general in charge of antitrust, to drop the investigation.[95]

Groves and Lansdale frequently found themselves at odds with other U.S. government agencies. This was a direct result of Groves's management philosophy of wanting to avoid unnecessarily informing anyone about the activities of the MED. For example, the Federal Bureau of Investigation was nominally in charge of general security, counterintelligence, and countersubversion in the United States. Yet Lansdale operated his security office across the United States since Groves had no intention of sharing the secrets of the atomic bomb with the Bureau.[96] Groves was even more vehement about keeping the U.S. atomic bomb program secret from the U.S. Department of State (long considered the government department least capable of keeping secrets from foreign powers—or anyone, for that matter).[97] He was concerned that in the course of performing his duties, he would be required to negotiate agreements with other nations, particularly Great Britain. This would apply if the United States decided to enter into a partnership with the British (which they did) to work together on scientific intelligence operations or on the acquisition of nuclear materials. Groves ordered Lansdale to prepare a legal memorandum providing Groves with the legal cover to use his executive branch authority to subvert the treaty process and engage in direct negotiations.[98]

By the winter of 1943–44, Groves had successfully consolidated his power, established an organizational foundation for future scientific intelligence efforts against Germany, tapped into the British scientific intelligence operation, slowed German progress toward its atomic ambitions, and shored up the security of the Manhattan Project (at least as far as the Germans were concerned—the Soviets were another matter altogether). There was still work to be done in each of these areas to achieve the perfection that Groves demanded (in himself and his subordinates). In fact, Groves would continue to dedicate time and attention to each throughout the remainder of the war. However, as Allied forces began to march up the Italian peninsula in the spring of 1944, it came time to put men in the field. It was time for the Alsos Mission.

3

Alsos

The Mission to Solve the Mystery of the German Bomb

John Lansdale later recalled that he first conceived the idea for a scientific intelligence mission to Europe "sometime around the middle of 1943." He believed that the only way to achieve an acceptable degree of certainty in regard to the German atomic bomb program was if the United States moved the "intelligence gathering activities into the very front line of the fighting activity." Since earlier methods of discovery had failed, Lansdale maintained "that there was no way to get such information except after the occupation of areas where the research was going on." Such a mission, he argued, could produce the "means of examining activities in universities, of sampling the water in various streams for radioactivity where the streams might have received discharge of cooling water from atomic piles and the like." In addition, Lansdale worried that German atomic research facilities, documents, and other key intelligence sources would be destroyed during and after battle. Retreating enemy forces were likely to carry away or destroy persons, documents, or equipment of possible value to the Allies. In addition, he feared the effects of the "inevitable looting by victorious front line troops." Victorious armies, he observed, frequently occupied large buildings, facilities, or enemy headquarters, scattering, damaging, or destroying scientific documents they did not and could not identify as important. Lansdale insisted that it was imperative that U.S. scientific intelligence get there first.[1]

That summer Lansdale brought his idea to Leslie Groves, who had already been thinking along those same lines. Together they formulated a general plan for a scientific intelligence mission to Italy, and presented it to Maj. Gen. George Strong, the G-2, explaining to him that, as Groves later wrote, this would be the best way to exploit "sources of information that would become available to [the United States] as the American Fifth Army advanced up the Italian peninsula."[2] With the concurrence and support of Vannevar Bush and the OSRD, Strong submitted the proposal for approval to Gen. George Marshall, the army chief of staff. In his memorandum, dated September 25, 1943, Strong told Marshall that "while the major portion of the enemy's secret scientific developments is being conducted in Germany, it is very likely that such valuable information can be obtained thereon by interviewing scientists in Italy." The proposal recommended that the mission be made up of a commanding officer (either a lieutenant colonel or colonel), no more than six interpreters (of various military grades), no more than six Counter Intelligence Corps Special Agents as investigators (also of various grades), and no more than six scientists (either civilian or military). According to Strong, "This group would form the nucleus for similar activity in other enemy and enemy occupied countries when circumstances permit."[3]

Lansdale, Groves, and Strong were acutely aware that if the scientific intelligence mission was seen as overtly targeting atomic scientists and installations, it could alert the Germans, causing them to redouble their atomic research efforts and push their program even further underground. Thus they decided to disguise the true purpose of the mission by broadening it to target all areas of German scientific research, not just atomic weapons. The mission, Groves explained, was directed "to exploit to the fullest sources in a number of fields of technical interest."[4] In Strong's memorandum to Marshall, he wrote, "The scope of inquiry should cover all principal scientific military developments and the investigations should be conducted in a manner to gain knowledge of enemy progress *without disclosing our interest in any particular field*" (emphasis added).[5] Groves was so intent on drawing "attention away from the mission's interest in atomic matters" that he decided that, on paper at least, the mission should report directly to Strong at G-2. Strong would then relay the information to the appropriate agency, atomic intelligence to Groves, and other scientific intelligence to whoever was most interested.[6]

Despite these precautions, Groves worried that the primary goals of the mission would be discovered by the enemy. Part of this fear stemmed from the name of the mission itself. The unnamed individual or individuals in G-2 who were tasked with assigning code names to operations decided to name the scientific intelligence mission "Alsos." Although it sounded innocuous enough, and in many cases it was assumed to be an obscure acronym, *alsos* is actually the Greek word for "grove" or "a grove of trees." Someone at G-2 with a misplaced sense of humor thought it would serve as an homage to the MED director. Terrified that the mission's secrecy would be compromised even before the mission began, Groves briefly contemplated ordering the name to be changed, but in the end decided that to change the name would only bring attention to the operation.[7]

Fielding the Mission

The next task was to choose the scientific personnel for the operation. This was left to the OSRD, Vannevar Bush, and Bush's deputy, Carroll L. Wilson. In late September, Wilson sent Bush a memorandum outlining the scientific goals of the mission, and recommending a number of scientists who were qualified to serve in Alsos. Describing the mission in scientific intelligence terms, Wilson wrote that "the purpose ostensibly, and in a very real sense, would be to send some scientific personnel familiar with important phases of the OSRD program to Italy for the purpose of interviewing Italian scientists, if and when available for such interviews, both to determine the current status of Italian research and development and to find out as much as possible through such individuals concerning German work." He then revealed the true reason for Alsos: "A very definite purpose of the Mission would be to find out information in the S-1 field ["S-1" was the American code for atomic bomb research] and presumably one or two of the scientific personnel, perhaps a physicist and a physical chemist, would be given sufficient information concerning the S-1 program here to allow them to probe intelligently for information in this field. Although this might be the true purpose of the Mission, it would be masked behind the façade of general scientific interests."[8]

Wilson's memorandum continued by listing the preferred traits and characteristics of the ideal scientist for a scientific intelligence operation. He argued that "the most desirable combination of qualifications would be fluency in Italian, acquaintance with Italian scientists, and full clearance and knowledge of important parts of the OSRD program." Unfortunately there were very few scientists who met these criteria. Most of those who were cleared for secret work did not have the Italian language skills necessary for the operation, and Wilson believed that interpreters alone would not suffice: "[The reason] for wishing to have a scientist who speaks Italian fluently is that much of the scientific and technical terminology would be unknown to an interpreter not trained in scientific work."[9] The one exception was Major Will Allis, an MIT-educated physicist who was on loan from the NRDC to the War Department. Allis had grown up in France; spoke French, German, and Italian fluently; and was fully acquainted with the OSRD's scientific programs, particularly the top-secret radar project. Major Allis was a perfect fit for Alsos.

The makeup of the rest of the scientific team, Wilson said, "would depend upon the fields in which [the mission is] most likely to discover information of value." In addition to atomic intelligence, of course, Wilson suggested that the group be composed of scientists who could exploit enemy developments in the fields of radar, communications, guided missiles, rockets, explosives and general chemical developments, and controlled torpedoes. His memorandum provided Bush with a list of potential candidates and their qualifications, "some personnel who would be available for consideration as members of such a Mission." The list included I. I. Rabi of Columbia University (a well-known physicist who was personally acquainted with some of the Italian physicists), Louis Turner of Princeton (a well-known physicist undoubtedly known to the various Italian physicists), David Langmuir of the OSRD (who was fully acquainted with both the U.S. and British radar programs), Samuel Goudsmit of the University of Michigan (a well-known nuclear physicist who knew some of the Italian physicists, although he did not speak Italian), Ralph E. Gibson of the Carnegie Institution (chair of the Rocket Propellant Panel of the Joint Committee on New Weapons and Equipment), Robert Shankland of the Case School for Applied Science in Cleveland, Ohio (head of the Underwater Sound Reference Labs), Alfred Murray of the NDRC (who in college had studied technical German for three years, and French for one and a half years), T. R. Hogness of the University of Chicago (who was connected with the chemical divisions of the OSRD for two years and also, for about a year or more, was a member of the London Mission of the OSRD, covering chemical developments), George Kistiakowsky of the NDRC (who was in charge of the explosives research and development work for the NDRC for two years and had directed the efforts of an extensive NRDC program on explosives and propellants—and who also undoubtedly knew some of the Italian chemists), and John Johnson of Cornell University (acting head of the London Mission, with particular responsibilities in the chemical field).[10]

Wilson's list of scientists provided Bush with everything he needed in order to assign what Bush called "top grade scientific personnel" to the Alsos Mission. In a letter to Groves, Bush recommended George Kistiakowsky as the best candidate for the operation, arguing that "if you have this man on the job you will need no one else, except for auxiliary purposes and scenery." He also concurred with the recommendation of Will Allis for the mission. "The suggestion of Major Allis as one of the officers," Bush wrote, "seems to me to be excellent from his knowledge and background. . . . I think he would make a useful individual to go along."[11] Rounding out the scientific team was James B. Fisk of Bell Laboratories. Fisk had been working with the OSRD under Bush and Wilson and could be temporarily spared from his scientific work for such an important operation.

Bush and Wilson had discussed the idea of including on the mission a representative from the National Advisory Committee for Aeronautics (NACA). The committee wanted to send one of its scientists to Italy in order to discover what it could about enemy aircraft developments. Both Bush and Wilson felt that this was a prudent suggestion. Wilson argued that the "addition of an aeronautical engineer would diversify the group and add further camouflage to the S-1 purpose,"[12] and Bush concurred: a NACA representative, he said, "certainly will provide means for disguising the objects of the mission when necessary."[13] In the end, however, Groves and Lansdale decided not to include a NACA scientist. Their reasons are not entirely clear from the available documents, but it seems likely that they felt the potential costs outweighed the benefits. A representative from NACA would help to hide the true intentions of the mission, yet extra members meant extra security risks, and NACA was not an organization whose support was essential to the overall mission.

The same could not be said about the Department of the Navy, which requested inclusion in the mission in November. Lieutenant Bruce Old, a classmate of Will Allis's at MIT and a scientist in the Office of the Coordinator of Research and Development in the Navy Department, had heard through Allis of the operation being assembled for the Italian theater. Old spoke to Lansdale, and with Groves's permission Lansdale asked Old to join the mission. In a memorandum to the Army G-2, Major General Strong, Lansdale explained the MED's rationale for including the navy in Alsos. Lansdale wrote to Strong that Groves was "of the opinion that minor Naval participation would be most desirable and that every effort should be made to arrange it so that any Naval participation would be under our control." In addition, Lansdale maintained that "it was believed that any Naval officer accompanying the expedition should be assigned to the detachment and be a part of it subject to the direction of the Commanding Officer thereof, and subject to the general supervision of Mr. Carroll Wilson, NDRC, with reference to the scientific mission."[14]

On November 10, 1943, Secretary of the Navy Frank Knox formally requested navy representation in Alsos. He wrote to Henry Stimson, the secretary of war, that Alsos "offers such interesting possibilities for obtaining valuable technical information."[15] Stimson agreed, and on November 16 officially approved Lieutenant Old as a member of the Alsos Mission.[16] Naval participation would help to mask the mission's true intentions, but more importantly the inclusion of Old would guarantee that Alsos would receive the full support of the entire U.S. military.[17]

All that remained was to choose the military commander of the mission, and for this Lansdale and Groves selected forty-three-year-old Lt. Col. Boris Pash. The son of a Russian Orthodox priest, Pash had been born with the name Boris Pashkovsky in California but had moved to Russia as a teenager. He had his first taste of combat fighting against the Bolsheviks during and after the Russian Revolution. Pash returned to the United States in the 1920s when it became clear that the Bolsheviks would maintain their control in Russia. He joined the Army Reserve and was called to duty in Army Intelligence in 1940 when the U.S. military began to mobilize. A bitter opponent of Communism, Pash was assigned the task of investigating Communist subversion in the San Francisco–Berkeley area, where, as Lansdale later wrote, his job required him to pay close attention to a number of "young scientists and technicians working in the many scientifically oriented establishments" who were active members of the Communist Party.[18] He first came to the attention of Lansdale and Groves when he became the liaison to the MED for the Western Defense Command, where, according to Groves, his "thorough competence and great drive had made a lasting impression" on the MED director.[19] By early 1943, Lansdale noted, Pash was "conducting a wide spread and complex investigation of communist activity" at the Berkeley Radiation Laboratory, following suspected party members (including Robert Oppenheimer) and bugging their homes and the places they frequented.[20]

Pash was a perfect fit for the organization Groves was trying to create. He was motivated, dedicated to the mission, and, perhaps most importantly, completely loyal to Groves. "I had had experience with General Groves while working on the Soviet espionage case," Pash later recalled. "We had always come to a speedy meeting of minds—and there had never been a question as to whose mind was met! The General knew how to get results. He never tolerated the staff gobbledegook and beating around the bush of which there was so much in Washington. He was exactly the kind of man to be depended upon in a national emergency."[21]

Pash was officially transferred from the Western Defense Command to the MED in late November 1943, but by then it was just a formality. He had been part of operational planning for Alsos since at least early October, and by the time of his official transfer the infrastructure for his command had been finalized. He would serve as the commanding officer of a detachment that included four scientists (Major Allis of the War Department, Lieutenant Commander Old of the Navy Department, and Fisk and John Johnson of the OSRD),[22] four interpreters, six Counter Intelligence Corps (CIC) officers, and an executive officer—Capt. W. B. Stanard—who would manage many of the administrative tasks of Alsos. The CIC officers would be assigned to the mission once it had established itself in theater.[23] For security purposes, of the scientists only Fisk was fully briefed on the primary focus of the mission: German work on an atomic bomb.[24]

According to Groves, the Alsos Mission was unlike any intelligence operation previously assembled: "Its make-up was considerably different from that of other intelligence units. It included people who were capable of extracting through interrogation and observation detailed scientific information on atomic energy. It also contained people who were generally familiar with the research programs and interests of both the United States and Great Britain and, insofar as possible, of our enemies. The members of the mission had to have general knowledge of enemy equipment and they had to be prepared to seek out not only military laboratories and technical personnel, but civilian scientists, technicians and facilities as well."[25]

On November 26, General Strong brought Pash to meet Secretary of War Stimson, who provided him with a letter of introduction to Gen. Dwight Eisenhower, the commander of Allied forces in North Africa and Italy. Stimson's letter identified Pash as an officer who "has been specifically charged with the procurement of information concerning the scientific activities and developments of the enemy." The letter informed Eisenhower that U.S. intelligence "believed that a large amount of such information is available within the territories under your command." The secretary of war considered the mission "to be of the highest importance," and added, "It is essential that it be accomplished expeditiously and successfully." Stimson asked Eisenhower to "give Colonel Pash every facility and assistance at your disposal which may be necessary or helpful in the speedy completion of this mission."[26] Pash was also given a letter from General Strong to Gen. W. Bedell Smith, Eisenhower's chief of staff. This letter, according to Pash, "contained a request that any needed personnel, equipment and funds be made available and that direct communication with Washington be arranged for me." This was a fairly significant departure from standard operating procedure, particularly in the case of someone as junior as a lieutenant colonel, and according to Pash, "The letters were a further indication of the importance Washington attached to the operation."[27]

Alsos Italy

On December 13, 1943, the Alsos Mission assembled in Algiers and Pash reported to Allied Force Headquarters (AFHQ) and General Smith.[28] Pash told Lansdale that after presenting Smith with the letters from

General Strong and the secretary of war, Pash told him "of the interest of the President in the Project [the MED]," and gave Smith "some general information pertaining to the project and to the aims of the mission as they relate to this project." Smith promised that the mission would receive every priority, and he gave Pash "verbal approval and orders to send reports and communications relating to the project without filing copies of such reports at AFHQ." Smith then told Pash that the deputy G-2, Colonel Roderick, would be his contact at the headquarters, and that Roderick had been instructed "to make all necessary arrangements for the mission."[29]

In the short term, "necessary arrangements" meant transportation. Roderick provided the Alsos Mission with seats on the first available flight to Naples, where they arrived on December 15, and Captain Stanard established their Alsos Mission headquarters at the Bank of Naples (Banco di Napoli). Smith had ordered them to report to the commander of the Allied Control Commission in Brindisi,[30] Maj. Gen. A. K. Joyce, after they had established themselves in Italy. However, once they arrived in Brindisi on the seventeenth they discovered that Joyce was temporarily away from headquarters, and so they reported to Joyce's deputy, Brig. Gen. Maxwell Taylor. Taylor had not been briefed on the mission, and according to Pash, "He flatly stated that he would not do a thing for us unless we told him the whole story." After some negotiation, Taylor agreed to have one of his officers attend to their immediate needs while they waited for the return of General Joyce, who was expected back the following afternoon.

In the meantime, Pash and the members of Alsos traveled to Taranto, where Lieutenant Commander Old had arranged a meeting with Captain Zaroli of the U.S. Navy and the Allied Navy Command.[31] Pash explained to Zaroli the general aspects of the mission, and subsequently told Lansdale that he believed Zaroli would "be very useful in establishing necessary contacts" in both the U.S. and Italian Navies.[32] In fact, Zaroli introduced the Alsos team to Lieutenant General Matteini, head of Italian Navy Ordnance, and other high officials of the Italian Navy.[33] After leaving Old and Allis in Taranto to continue liaison with the navies, Pash, Fisk, and Johnson returned to Brindisi to meet with General Joyce. Over an hour was spent with Joyce, "during which time," Pash reported to Lansdale, they "went into more detail about the mission." After this explanation, Pash said, Joyce was "extremely cooperative and said that he will do everything to have the mission succeed." He called in General Taylor, explained to him the importance of the mission, and told him of "the need and desire to [give] priority [to] all [Alsos's] requests." According to Pash, "It was the first encouraging reaction we [received] since we left Washington, beside the attitude of Gen. Smith, CofS, AFHQ." After the mission was explained to both generals, Alsos "received 100% cooperation," Pash reported, adding, "And it is my opinion that no other headquarters and no other officer could give any more support than was received by both generals." Both Joyce and Taylor, Pash told Lansdale, promised "that any request made by [Alsos] and which they are in position to grant will be given first priority. Subsequent events have indicated that they meant what they said."[34]

"Subsequent events" included a meeting, arranged by Joyce, between Alsos scientists and the Italian minister of communications, which took place on December 20. The minister promised to introduce the Americans to prominent Italians who might know something about German scientific progress. Most importantly, Joyce assisted in brokering a meeting on December 23 between Pash and Marshal Pietro Badoglio, head of the Italian provisional government and soon-to-be Italian prime minister. Badoglio provided Pash with yet another letter of introduction, this time addressed to all Italian civilian and military authorities.[35]

On the twenty-fifth, all the members of Alsos reconvened in Naples. During the week in Italy, Pash had established relationships with the appropriate authorities to smooth the way for Alsos operations, allowing "the mission to become substantially independent of any formal organizations in this theater," the report indicated. The scientists of the mission had spent the week interviewing "all of the available informed individuals who have information of special interest." Intelligence had been collected "regarding some German developments as well as a fairly complete picture of Italian research efforts and results." While there was doubtless "further information available which [would] form an important background for activities in Rome," the report said, the scientists believed it was evident "that the information of importance which [had] been available in Southern Italy [had by then] been given either to members of [the] mission or, during the past three months, to other intelligence units."[36]

The Italian scientists most likely to have information about the German atomic bomb program were Edoardo Amaldi and Gian Carlo Wick, both of whom were believed to be in Rome. Wick and Amaldi were nuclear physicists who had been close collaborators of Enrico Fermi's before Fermi left for the United States. They also both knew Werner Heisenberg well, and could perhaps provide Alsos with key information about the most important of German atomic scientists. On December 28, Pash visited the headquarters of the military unit tasked with the capture of Rome, the U.S. Fifth Army. There he met with a Colonel Howard, the Fifth Army's G-2, and, according to Pash, "established very satisfactory relationships" with the G-2 "and associated sections of the Fifth Army."[37] Together with Colonel Howard and the commander of the Fifth Army, Lt. Gen, Mark Clark, Pash developed the plan for the Alsos Mission's operations in Rome. They decided that the advanced force of the Alsos Mission (Pash, Old, Allis, and the CIC agents—the military personnel) "will go into Rome with first occupying forces." The mission would be "to secure all scientific documents and pick up such people as may be of value," and to prevent their dispersal or destruction. The scientists of Alsos would follow when the battlefield was secured.[38]

The plan in place, all that remained was for Allied forces to break through German resistance and get to Rome. Yet the Allied armies had been slow to reach this goal, hampered by bad weather, rough terrain that favored the defenders, and strong German resistance. The advance up the Italian peninsula was halted short of Rome at what was known as the Gustav Line, a series of German defensive fortifications that ran from coast to coast across Italy. During the first weeks of January, Pash and members of Alsos made frequent trips to Fifth Army headquarters to inquire about the status of the war. What they heard, Pash later wrote, was "discouraging as far as [our] mission was concerned. The campaign was in a static period and there was no hope of reaching Rome soon."[39] Pash was getting more and more frustrated. The scientists were keeping busy interviewing any Italian scientists and captured technical specialists, "but those early contacts," he recalled, "indicated that no startling results could be expected even though some valuable scientific intelligence [not atomic bomb–related] was being picked up."[40]

There was so little for Alsos to do that Pash decided, in consultation with Fisk, to send Johnson back to the United States, "in view of the fact that the . . . situation does not justify retaining both him and Fisk," he reported to Lansdale. Pash also decided that if the Fifth Army did not break through to Rome in the very near future, he planned on sending Fisk home as well, arguing that "it seems that a mission of this nature does not require men of Fisk's or Johnson's calibre." Old and Allis were performing well and could handle the scientific tasks alone. Highly qualified scientists such as Fisk and Johnson, Pash said, were being wasted in Italy, particularly "when they are kept unoccupied for a period of time due to lack of available personnel for questioning. . . . This is the situation in our case."[41]

Pash regained some hope in the third week of January when he heard about the planned Allied amphibious landing against German forces in the area of Anzio. Code-named Operation Shingle, the attack on Anzio was intended to outflank the German forces of the Gustav Line and open an alternate route to Rome. Buoyed by the chance of a breakthrough, Pash began planning in earnest:

> We have evolved a plan in connection with the probable demonstration against Rome. . . . Our party has been divided into two groups. The forward echelon will be commanded by me and we will go on to Rome with what is known as the S Force. This will be an amphibious operation, and by the time you get the report I'll either be in Rome or will be dead, or maybe both. Anyway, we will go in to secure as rapidly as possible the objectives (buildings and persons) important to our mission. Old and Allis will go with me. Dr. Fisk, Capt. Stanard and three agents will come up by land with most of our equipment. This plan seemed the most practicable one, if we were to get to Rome in a hurry and secure targets of interest to us before they were destroyed or made unavailable.[42]

The Anzio operation, however, did not achieve its ambitious goals. The Allies established a beachhead, but indecision and inaction by the mission commander, Maj. Gen. John Lucas, prevented the invading forces from exploiting the advantage of surprise. The delay allowed the Germans to surround the Allied force, and only the most heroic efforts allowed the Fifth Army to hold on to its foothold at Anzio. Clearly it would be some time before the Allies, and Alsos, entered Rome. Yet the men of U.S. atomic intelligence were not content to remain idle while military operations floundered. If Groves, Lansdale, Furman, and Pash could not get to

Wick and Amaldi through a general advance, covert actions were necessary. This meant calling in the OSS.

The OSS began working on the German atomic intelligence problem in the fall of 1943. From early November through late December, Donovan's appointee as the head of the Technical Section of the Secret Intelligence Branch, Col. Howard Dix, sent a series of requests for information on Italian and German atomic scientists to Allen Dulles, the top OSS official in Bern, Switzerland. Dulles was asked to provide the locations of thirty-three scientists, three of whom were Italian (Wick and Amaldi among them) and thirty of whom were German. The names of the scientists were coded: Werner Heisenberg was "Christopher," Otto Hahn was "Tag," Carl von Weizsäcker was "Lender," Wolfgang Gentner was "Ernst," and so on.[43] When Dulles had intelligence to provide, he sent it to Dix under the code designation "Azusa," indicating that it was atomic intelligence and ensuring that it was promptly brought to the attention of Groves and the MED intelligence team.

Dulles cultivated a number of intelligence sources in Switzerland, but none more important to the Americans' understanding of the German bomb program than fifty-three-year-old Paul Scherrer. Scherrer was a Swiss physicist and professor at the Federal Technical College in Zurich. He was not a scientist of the same order as Heisenberg, Hahn, or most of the other top German atomic scientists. Yet Scherrer had been attending many of the same academic conferences as they had since the early 1920s, and as a result knew most of them rather well and had been close friends with some of them, most notably Heisenberg, for over two decades. Code-named Flute by the OSS, Scherrer was never a formal agent of the OSS (he was never paid or formally recognized for his contributions). He did all he could, however, to aid the Allied cause, and in so doing provided Dulles, the OSS, and the MED intelligence team with arguably the most productive insight into German atomic progress. In the spring of 1944, Scherrer gave Dulles what could essentially be called a bomb damage assessment of the direct attacks on German scientific facilities and scientists ordered in the fall of 1943 by Leslie Groves. He told him that the Kaiser Wilhelm Institute for Chemistry outside Berlin (Otto Hahn's institute) had been partially destroyed, and that the scientific institutes at Munich, Leipzig, and Cologne had been damaged beyond immediate repair. While Heisenberg's institute, the Kaiser Wilhelm Institute for Physics, had remained untouched, Scherrer told the OSS that the Germans were busy constructing alternate facilities in the countryside, where they would move their prominent scientists to protect them from Allied attacks. Some of those scientists, according to Scherrer, had already begun to disperse throughout rural Germany; Weizsäcker and others had reportedly moved to Strasbourg, in the Alsace region. In southern Germany, in the vicinity of the towns of Bissingen and Hechingen, the Germans were building a laboratory with a two-hundred-million-volt cyclotron, presumably for eventual use by Heisenberg and Hahn.[44]

While Dulles was busy exploiting sources in Europe, the OSS and Groves were planning covert operations back in Washington. One such operation, concocted in late 1943, was code-named Project Larson, and was designed to have an OSS agent infiltrate occupied Italy to interview Italian scientists about German atomic research. Chosen for the mission was the forty-one-year-old OSS agent Morris "Moe" Berg.

Moe Berg was a professional baseball player before he was a spy. His baseball career began in 1923 when he was signed as a shortstop for the Brooklyn Robins (later Dodgers) of the National League. After hitting only .186 for Brooklyn (zero home runs and only six RBIs in almost fifty games), he was sent to the minor leagues, where he languished until 1926 when he signed with the Chicago White Sox. The White Sox would eventually move Berg to catcher, and although his lifetime statistics were mediocre at best (.243 average with only six career home runs), this positional change would allow Berg to remain steadily employed in Major League Baseball for fifteen years (his understanding of the game, his knowledge of hitters,[45] and his defensive skills made him a valued asset, despite his offensive liabilities). Berg moved around the league throughout his career, leaving Chicago for Cleveland and Cleveland for a job with the Washington Senators, then going back to Cleveland for a year before finally settling in to finish his baseball life with the Boston Red Sox. From there he would retire in 1942 at the age of forty.[46]

Berg's intellectual acumen made him a legend in the intelligence community.[47] He attended Princeton University, where he studied modern languages, graduating with fluency in Latin, Greek, French, Spanish, Italian, German, and Sanskrit (during his life Berg allegedly learned as many as twelve languages). After graduating, Berg studied French at the Sorbonne in Paris, and during the baseball off-season attended Columbia

Law School, where he earned his law degree in 1928. As a member of the Washington Senators in 1934, Berg was included on a team of all-stars, including Babe Ruth and Lou Gehrig, who went to Japan on a goodwill mission to play the Japanese all-stars. Since Berg was a light-hitting third-string catcher, it surprised many baseball insiders that he was asked to join the team.

After giving an eloquent speech on Japanese-American relations at Meiji University, however, Berg set out to complete a secret intelligence mission. Equipped with a camera, Berg snuck to the roof of a Tokyo hospital and took pictures of Tokyo Harbor, naval installations, and other high-value military targets. In 1942, the pilots of the Doolittle Raid analyzed these same photos before their famous bombing run, although most of the photos were too old to be of much use. When the Second World War began, Berg volunteered for service, and he was assigned to the Office of Inter-American Affairs. He was sent to Latin America in 1942, where he used his fluency in Spanish to persuade government officials, journalists, and businessmen to resist joining the Axis cause. In 1943 Berg was recruited by William Donovan and the OSS, and they immediately put him to use, dropping him by parachute into occupied Yugoslavia. There he met with both opposition forces to assess their strengths and to recommend to the United States which group should be supported. After meeting with King Peter's Chetniks and Tito's Partisans, Berg concluded that Tito was better equipped to fight the Nazis, and U.S. aid thus went to Tito.[48]

When Berg was assigned to Project Larson in late 1943, his task was to sneak into Rome to interview physicists at the University of Rome about the German atomic bomb project, and try to discover the whereabouts of its supposed leaders—Heisenberg, Hahn, Weizsäcker, and the rest. Unfortunately, he would not get the chance to accomplish this mission until the following summer. Gen. Mark Clark, commander of the Fifth Army and no great fan of the OSS, refused to allow Berg to enter the theater. While waiting for permission to enter Italy, Berg took the time to teach himself quantum theory and matrix mechanics. He read the German physicist Max Born's *Experiment and Theory in Physics*,[49] and he studied Heisenberg and his famous Uncertainty Principle. By no means an expert, Berg had taught himself enough of the physics of the atomic bomb to understand what it would take to successfully build one. This knowledge would be an invaluable asset in the coming months.

Introspection and Reorganization

When it became evident in early 1944 that the Allies would not break the German lines and enter Rome for some time, Groves decided to redeploy the Alsos Mission back to the United States. The first to return was Fisk in early February. Bell Laboratories had written Vannevar Bush, arguing that since Fisk was not being effectively utilized in Italy, he should be immediately reassigned. Bush wrote Groves and told him, "I feel that unless there is urgent reason for his remaining in Italy, he should perhaps return to this country."[50] By the end of February, Pash and Allis redeployed to the United States, and by the first week of March the remaining members, including Old and Stanard, were back in Washington. The CIC agents, borrowed from other U.S. Army forces in the Italian theater, were sent back to their home units on the understanding that the same personnel would be reassigned again to Alsos if and when the mission was resumed.[51]

Upon their return to the United States, the members of the mission were asked to produce written reports on the Italian operation. Fisk, the first back in the United States, was the first to do so. Fisk was also the only member of Alsos outside of Pash who knew the true objectives of the mission. Thus his reports (he wrote two) included both atomic intelligence and general Italian and German scientific developments. As far as the general science was concerned, Fisk's report of February 14 was essentially a summarization of a January 22, 1944, report compiled by Fisk, Allis, Old, and Johnson.[52] It stated that the Alsos Mission had discovered information that could be valuable to Allied forces on a number of topics, including rockets, ordnance, guided missiles, fire control, explosives, chemical weapons, communications, radar, and infrared.[53]

In the atomic field, Fisk's report of February 5 provided information garnered from Italian contacts established by Alsos. He wrote that during attempts to "obtain direct evidence" on German atomic research, "it was unnecessary to use any great subtlety [with the Italian scientists] and it never became necessary to reveal our interest in the matter. Without exception the individuals approached were anxious to be of assistance and without exception they informed us that in all war research the Germans had been most secretive." Therefore, Fisk said, actionable intelligence was at a premium: "Hence any evidence which may be of interest amongst the following fragments will be indirect and for the most part negative."

Fisk explained that Alsos had also tried to "build up a general picture of German war research activity and industrial activity which might subsequently allow a reasonable deduction of their interest and progress" in the atomic field.[54]

Sources of information for Alsos included (among others) a Professor Wolfers of Algiers; a Professor Henriot, an exiled Belgian physicist Alsos interviewed in Algiers; a Professor Calosi, who had worked with the Germans and prominent Italian scientists in Rome before returning to Naples; and a Professor Tiberio, a former student and collaborator of Edoardo Amaldi's in Rome. Wolfers and Henriot told Alsos that several prominent French physicists were tortured to death by the Germans, "presumably for refusal to reveal some scientific knowledge," according to Fisk. Wolfers reported rumors he had heard that the Paris cyclotron had been moved to Germany, and Henriot stated that Frédéric Joliot-Curie was in Paris and might be working on fission (the two reports seem mutually exclusive, but that did not appear to faze the scientists). He also said he did not think the Germans were working on an atomic bomb, but Fisk believed Henriot "does not know for he had no evidence one way or the other." Tiberio was reported to have told Alsos that he had talked to Amaldi in Rome in June 1943, and that Amaldi had told him "that Germans have tried nuclear explosive [*sic*] but have not succeeded." Finally, Fisk reported, Calosi stated "that the Germans had said a number of times that only those things which were in hand at the beginning of the war would be of any material use in the prosecution and final outcome of the war." Calosi felt that this was the German "guiding philosophy," and he did not think "the Germans had in the course of development any 'fantastic new weapon.'"[55]

This was the extent of atomic intelligence collected by the Alsos Mission in Italy. Making matters worse, in an exit interview with Robert Furman, Fisk warned that even this meager information should be considered suspect. He told Furman that Henriot got his intelligence from the scientific "grapevine," and that he felt Henriot "was not in possession of much recent information." Fisk also thought Henriot was "not a good judge of what might be going on" since he had preconceived ideas about the impossibility of atomic weapons. He argued that Wolfers was not a serious source of information since it was most likely the case that Wolfers was just passing along intelligence that "probably comes from Henriot."

Fisk then indicated that the scientists interviewed were not considered first rate (they were "not of the Fermi school"), and that the University of Naples, where Professors Calosi and Tiberio worked, was "definitely a second rate university according to American standards." In all, Fisk told Furman that until Alsos entered Rome, very little concerning the German atomic program would be discovered, and perhaps not even then: "No one was found in southern Italy who was fundamentally interested in fission research. No one was studying the literature thoroughly. No one had written back to the German scientists who were writing on fission to question their thoughts." According to Fisk, "An enormous barrier exists between the Italians and Germans as to the war effort."[56]

Fisk believed the Germans were wise to exclude the Italians from atomic research "as unreliable." The Italians in the South could not make a viable contribution to the German atomic research effort. Fisk noted that the Italians "never expressed the thought that something ought to be done about fission during the war," and that the idea of an atomic bomb "was looked upon as fantastic and inconsequential." Fisk described to Furman a meeting with Italian scientists in which one scientist stopped the conversation in order to inform other (supposedly qualified) physicists "how fission would work in a bomb." Unfortunately, Fisk said, "his explanation showed he was not very well informed."[57]

Despite the paucity of intelligence on the German atomic bomb program, however, Groves and Lansdale were not discouraged. From the first suggestion of a scientific mission to Europe in the summer of 1943, they understood that it was unlikely that any information of value would be available in Italy. Alsos was intended as a dress rehearsal or prototype for later missions in France and Germany, or, in Lansdale's words, as a "training process."[58] Both he and Groves perceived Alsos in Italy "as a unique opportunity to give [Alsos] a dry run or exercise to prepare it for the effort to acquire information about the German atomic program after the Allied landings in Europe."[59] By this standard, the Alsos Mission in Italy was, to Groves and others, "most successful," Groves wrote.[60] It had shown that this type of operation, never before attempted, was feasible. Its success in working in the field, establishing the necessary contacts, and exploiting intelligence sources demonstrated that the members of Alsos would be able to collect valuable atomic intelligence when it became available to them in the future.

In addition, Alsos had done enough in other scientific fields to convince the scientific and military leadership that it was a worthy program. On February 29, Vannevar Bush recommended to Groves that based on the mission's results, Alsos should be continued. Bush felt "that this has been a decidedly interesting experiment, and although the specific results of interest to your project have been few, some of the information obtained by the Mission which relates to work of the NDRC has been most significant and one or two items have, in my opinion, justified the whole enterprise."[61] Bush's executive assistant at the OSRD, Carroll Wilson, also believed that "although the results were rather meagre as far as your particular interests were concerned, both the direct and indirect results in other fields certainly repay the effort involved in organizing and conducting the Mission."[62] Not surprisingly, Furman and Pash agreed; each recommended the continuation of Alsos in memorandums to Groves on March 6.[63] Furman argued that Alsos was so important that even if the scientific and military hierarchy decided to end their sponsorship of the mission, "the continuance of the mission behind the invasion forces entering Rome and the organization of a mission to go behind the Allies' invasion of Europe should be undertaken by this office if the efforts of Col. Pash to secure the support of other offices for this mission fail."[64]

On the basis of Bush's recommendation, the experience Alsos gained in Italy, and the recommendations of Lansdale, Furman, Pash, and the rest of his team, Groves wrote a memorandum on March 10, 1944, to the new assistant chief of staff, G-2, Maj. Gen. Clayton Bissell,[65] requesting the continuation of the Alsos Mission. Groves argued that the mission had withdrawn "from the Mediterranean Theater after completing its objective insofar as the situation in the Theater permitted." He explained to Bissell that "the presence of specially trained and unusually qualified specialists proved to be of positive assistance to the regular G-2 agencies, who took advantage of the ability of the technical personnel to make a proper scientific evaluation of available information." After briefly detailing Alsos's nonatomic scientific discoveries, Groves recommended that "the Alsos Mission should continue its present plan of operations in Italy," including a "prompt entry into Rome when it falls under Allied control to secure individuals and documents." Finally, Groves argued that "a similar scientific mission with the same general objectives should be made ready for use in other European territory as soon as the progress of the war permits."[66]

Three weeks later, Bissell wrote to George Marshall recommending the organization of Alsos for scientific intelligence on a permanent basis. Bissell believed the "high value of the recent scientific intelligence mission" demonstrated that the program should be continued in other theaters in a similar manner. Although the invasion of Western Europe was not imminent, Bissell argued that since opportunities for scientific intelligence rapidly disappeared on the battlefield (due to destruction by retreating forces, looting, etc.), it was imperative that the mission be organized in advance and "held in readiness." This meant that personnel would need to be permanently assigned to the mission in order to work efficiently at a moment's notice.[67] The reorganized mission would require new scientific personnel, ideally scientists who could remain with the mission for the duration of the war. These would be selected by Groves and Bush, and any new intelligence and administrative personnel would be chosen and furnished by G-2.[68]

The delay between Groves's request for the continuation of the Alsos Mission and Bissell's letter to Marshall was caused by bureaucratic disputes, not any question as to the usefulness of the mission. During March, several members of the Army General Staff proposed to Marshall a plan to centralize all scientific and technical intelligence under one organizational command. While this new umbrella organization would have brought together in one place all the units concerned with this very specific type of intelligence, it would have effectively *decentralized* atomic intelligence. Leslie Groves would no longer have had immediate control of atomic intelligence operations, which would have instead become one component of a broader scientific and technological intelligence effort.[69] However, the chief of staff and the secretary of war understood that the existing structure was well suited to the effective collection of atomic intelligence, and so on April 4, Marshall and Secretary Stimson approved the request for maintaining Alsos as an independent organization.[70]

The following day, Furman informed Groves of Marshall's approval, and told him that "a similar organization with the same general objectives will be made ready for use in other European territories immediately and sent to an active theatre as soon as the progress of the war permits."[71] Groves, Lansdale, and Furman had already begun planning in anticipation of approval, and in this they were joined by the deputy assistant chief of staff, G-2, Col. John Weckerling, Bissell's point man for Alsos. Weckerling would be in charge of assigning the military intelligence personnel for the mission, and on April 8 he designated Col. C. P. Nicholas to represent G-2 in the day-to-day supervision of the project. Nicholas had worked with MED intelligence in the past, and specifically Robert Furman during parts of the Italy mission.[72] That same day, Weckerling, in Bissell's name, informed the Navy Department and the OSRD of Marshall's decision and officially invited their participation in the next incarnation of Alsos.[73] For the OSRD and its chair, Vannevar Bush, this was merely a formality. Bush was heavily invested in the mission's success and would do whatever was necessary to assist Alsos. But the navy took almost a month of deliberation before it agreed to assign a member to the new mission on May 6. The navy decided to lend its support only after Commander Old demonstrated the value of Alsos to the navy leadership.[74]

In the meantime, Groves and the MED intelligence leadership had other problems. Several members of the Alsos Italy scientific team, most notably Fisk, had argued against retaining Boris Pash as military commander of the mission. According to Furman, Fisk told him that Pash had "no understanding of the scientific part of the mission"; Fisk believed the commanding officer "should have a broad understanding of the various fields of scientific activity which could be based upon engineering or scientific education." This in itself did not bother Groves, for Pash was not chosen as commanding officer for his scientific knowledge, but instead for his intelligence experience, aggressiveness in the field, and loyalty to Groves. More problematic was Fisk's other criticism of Pash, which directly questioned his leadership. Fisk argued that Pash abused the power he had been granted by the letters of introduction he was given by Stimson, Strong, and the Italian prime minister. He believed that Pash "exhibited these credentials unnecessarily at times which caused embarrassment to those in the mission and showed lack of good judgment." Furthermore, while Fisk acknowledged Pash's "persevering drive" and "enthusiasm" for his job, it was "this enthusiasm plus lack of real understanding as to the scientific objectives that caused him to make some errors . . . where excitement was certainly to be gained but likelihood of getting knowledge of enemy activities was rather remote. The risk involved in taking these chances did not appear to be worth the little knowledge that could be obtained."[75]

Major Allis agreed with much of what Fisk said, particularly the recommendation that the commanding officer of Alsos should be a scientist. Commander Old "concurred generally in this thought," while Johnson argued that personal relations were highly important for the success of the mission and that the commanding officer should have "the quality of congeniality" in order to relate to the mission's scientists.[76]

As a result of these comments, combined with what Furman called "cautiousness or perhaps suspicion on the part of Colonel Weckerling toward Lt. Col. Pash," Colonel Nicholas decided to proceed "carefully before accepting Col. Pash as Commanding Officer." According to Furman, "While the remarks made by the scientists were made in an unofficial manner and meant to be harmful, the appointment of Pash was nearly blocked." In reality, General Groves could have made a quick phone call and Pash would have been instantly reappointed, but Furman opted not to ask Groves to make this call and instead waited to see what Nicholas would decide. In the end, Nicholas reinstated Pash, with the caveat that the new mission organization would have, as Furman put it, "a more defined latitude of operation for the Commanding Officer who will confine himself to implementing and facilitating the plans" of a head scientist. The dean of scientists, who would later be officially called the chief of scientists, would be "responsible for carrying out the scientific investigations and making the necessary reports." In essence, the scientist would tell Pash where he needed to go, Pash would get him there, and then the scientist would investigate scientifically.[77]

It is possible that Nicholas was swayed in his decision to keep Pash as commanding officer by a memorandum Pash sent him in early April. Whether Furman told Pash about the scientists' comments or Pash correctly discerned their attitudes is uncertain, but Pash provided Nicholas with recommendations for the future Alsos operations that effectively answered many of his concerns. He argued that the "scientific members of the mission should control, to the extent that tactical conditions permit, the type of information sought and the selection of places in which and the persons from which the information should be obtained." Pash recommended that a member of the scientific group be designated as the "senior member of the group who will coordinate the activities of the group and whose decision will be accepted as final." This head scientist would "convey to the commanding officer of the mission the needs of the scientific group," and all requests for action "will come either through him to the commanding officer or will be called to his attention by the commanding officer."[78]

By the beginning of May, the organizational infrastructure of the new Alsos Mission was starting to take shape. Alsos would have a commanding officer (Pash), a chief scientist, and a newly established informal Advisory Committee that would assist in creating an overall intelligence plan, coordinate requests for information from other governmental agencies, and facilitate the movements of Alsos throughout the European theater. The Advisory Committee would consist of representatives of the director of naval intelligence, the director of the OSRD, the commanding general of the Army Service Forces, and the assistant chief of staff, G-2. This committee would concern itself with scientific intelligence of the nonatomic variety.[79]

The mission itself, Bissell wrote to Marshall, was designed to "follow the advance of Allied forces into occupied territory, remaining the necessary time after the enemy's defeat and making necessary visits and contacts in order to collect intelligence of the enemy's scientific developments." It would comprise two groups, independent but working together to accomplish the overall mission. A military and administrative group, consisting of the commanding officer, his executive (who acted as an administrative assistant), and the mission's interpreters, would be joined by a scientific group, consisting of the scientific chief (a civilian scientist) and "such additional military and civilian scientists as are attached to the mission with G-2 concurrence by the Director, OSRD, the Commanding General, Army Service Forces, and the Director of Naval Intelligence."[80]

As far as the day-to-day operations of the mission were concerned, the position of scientific chief was the only real departure from previous practices in Italy. This individual would have four main responsibilities: (1) to create the general plan in all its scientific aspects with regard to both objectives and personnel; (2) to prioritize the objectives with the assistance of the OSRD, the members of the first Alsos Mission, and the army and navy scientists on the current mission; (3) to evaluate the reliability and importance of intelligence sources; and (4) to determine the most effective approach to sources. The office of the scientific chief would also be tasked with keeping complete records of the activities of the mission, and with maintaining control of all collected scientific intelligence.[81]

On May 15, 1944, the forty-one-year-old nuclear physicist Samuel A. Goudsmit was named by the OSRD as the scientific chief of the Alsos Mission. Goudsmit was a Dutch-born, naturalized American physicist who had earned his PhD in the Netherlands under the world-renowned scientist Paul Ehrenfest. After he received his doctorate, Goudsmit moved to the United States and took a position as a professor of physics at the University of Michigan. When the war began, Goudsmit was recruited by the OSRD and the Radiation Laboratory at MIT to direct U.S. radar research. He was a candidate for the Alsos Mission to Italy, and his qualifications made him a natural fit for scientific chief.

For one thing, he was an exceptional scientist who had been on the cutting edge of nuclear physics for almost two decades. Since he had spent his early years in Europe, he spoke a number of languages fluently, including Dutch, French, and German. Also because of his Dutch ancestry and education, Goudsmit personally knew most of the French and German scientists whose research Alsos was tasked to investigate. He was also highly motivated. His parents, whom he had last seen in 1938 before leaving Europe for the United States for the final time before the war, were killed by the Germans in the Holocaust. Most important, however (at least as far as Groves and MED intelligence were concerned), was the simple fact that Goudsmit was not in any way involved in the Manhattan Project. He understood the principles of atomic fission, but he knew nothing about the progress of the U.S. bomb program, and therefore could not give away anything to the enemy if he were captured. Goudsmit himself said he was "expendable."[82]

The day of his official appointment as Alsos scientific chief, Goudsmit sent Colonel Nicholas a memorandum describing what he envisioned his job to be. Goudsmit wrote that "the purpose of Scientific Intelligence is to obtain knowledge about scientific war research in enemy and enemy occupied territory." The mission of Alsos should be limited to war equipment in the early stages of research, and "does not include information about enemy equipment which is already in use." To make this operation a success, Goudsmit said, "it is necessary to gather information about the location of research workers in enemy territory," and to discover intelligence "about the research laboratories of large industries as well as educational institutions." Goudsmit also stressed the importance of investigating German scientific publications for clues "on what types of research are not considered secret," and "what kind of university courses and research investigations receive special emphasis in enemy territory."[83]

In all, Goudsmit's ideas about the purpose and operational philosophy of Alsos were consistent with those of Vannevar Bush and the MED intelligence team. In fact, the relationship that would eventually develop between Goudsmit and Pash was much more amicable and, as a result, much more effective than the relationship between Pash and the scientists of the first Alsos Mission. The two became close friends, and continued their friendship until Goudsmit's death in the late 1970s. What made this relationship work when the others did not cannot be known for certain. Perhaps Pash evolved in his view of the utility of scientists on the battlefield, and no longer considered them "idiosyncratic 'longhairs'" whom he would have to lead "by the hand to keep them from blundering into trouble."[84] Perhaps Pash knew how close he had come to being replaced on the mission and made every attempt to stay in Goudsmit's good graces. Perhaps Goudsmit was just better equipped to handle Pash's own shortcomings. At any rate, the two got along well, and according to Goudsmit after the war, "Almost from the very beginning of operations in France, a clear understanding was reached concerning the division of responsibility between the military and the scientific groups of the Alsos Mission." He believed the arrangement between himself and Pash "worked out perfectly. Never did the military group question the judgment of the scientific group as to the importance of a target, and never did they fail to execute the operations as needed and planned." He concluded, "The Alsos method, it must be emphasized, succeeded only because of the close cooperation and mutual trust of the military and the scientists."[85]

In mid-May, at about the same time Goudsmit was receiving his appointment, Pash traveled to Great Britain, where he established the London office of the Alsos Mission. It was in London that Pash met for the first time one of the key officers of the MED intelligence team, Tony Calvert, who had been supplying him and the mission with information throughout the Italian campaign. Pash was impressed with Calvert, finding him "sharp, intelligent and at ease in any situation." Together, Pash and Calvert spent several days formulating the plans for the Alsos Mission's move to the Continent.[86] While in London, Pash also reported to Lt. Gen. Bedell Smith, who had been brought to Europe from Algiers to become the chief of staff, European Theater of Operations

(ETO). As in the Italian mission, Pash had been supplied with a letter from Secretary Stimson to General Eisenhower, commander of the ETO, asking for the general's full assistance for Alsos. Pash believed Eisenhower's and Smith's support would be imperative for the mission to succeed since the Alsos Mission was, in essence, a "bastard unit"—under Washington for operational control, but reliant on the ETO for administrative and logistical support.[87] This situation could have been a nightmare for Pash and Alsos, but because of Smith's familiarity with the mission and his strong belief in the necessity of its objectives, he did all he could to facilitate Alsos's success.

On this trip Pash would also contact Brig. Gen. Royal B. Lord (the deputy chief of staff, ETO), Brig. Gen. Edwin L. Sibert (G-2 of the First U.S. Army Group), Brig. Gen. T. J. Betts (G-2 of Supreme Headquarters Allied Expeditionary Force, or SHAEF), and the assistant chief of staff, G-2, ETO, Col. Bryan Conrad. Conrad, whom Pash described as "broad-minded, aggressive," and "good humored,"[88] would be "extremely helpful" in achieving the creation and maintenance of Alsos contacts with other U.S. government and U.S. military officials. According to Pash, every request submitted to Conrad was quickly granted, and "his able and willing assistance was instrumental in the accomplishment of the initial phase of the Mission's activity." Later, in early June, when the mission began planning for the shift of its operations across the channel to France, Conrad was so helpful to the mission's movements and logistics that Pash would call him "my guardian angel."[89]

Yet the mission to France would have to wait. On June 4, 1944, Rome finally fell to the U.S. Fifth Army, and Pash immediately traveled to Rome to take advantage of the breakthrough. Following directly behind the forward combat elements, Alsos moved into Rome at eight in the morning on June 5. The battlefield was still insecure (the Germans still controlled the northern half of Rome), but Pash and Alsos were able to make it through to Edoardo Amaldi's personal residence. After taking custody of Amaldi, Pash later wrote, Pash and Amaldi "talked about American scientists he knew and about whom I had been briefed."[90]

The Alsos Mission at this time, however, had severe limitations. Only Pash and the military element were actually in Italy, and it would take the mission two weeks before it could put a scientist in Rome. Alsos had established a foothold in the Italian capital, and Pash was able to take control of the appropriate Italian laboratories and scientific facilities, but no Alsos member then in Rome had the scientific credentials to exploit the information available.[91]

Fortunately, Moe Berg did. Robert Furman had the foresight to send him to London in May with the orders to move on to Rome when the situation allowed. Furman provided him with a list of prominent Italian scientists the MED wished him to interview, and on June 6 he began his mission in the home of Amaldi. Berg also managed to interview Gian Carlo Wick, and in a June 12 cable to OSS headquarters, he relayed his findings. Neither Amaldi nor Wick could provide direct information about the progress of the German atomic bomb program, but Wick did give Berg, the OSS, and the MED intelligence team valuable information about the locations and activities of some of the German atomic scientists, particularly Heisenberg. Wick, a former student of Heisenberg's, had kept in close correspondence with the German physicist throughout the war. He showed Berg a letter he had received from Heisenberg dated January of that year, in which Heisenberg revealed that his laboratories had been moved to a "woody region" in the "southern part of Germany."[92] While this was highly nonspecific, it did confirm what the Americans had learned from Paul Scherrer, and brought Alsos a step closer to reaching the crown jewel of its scientific intelligence mission.

While Berg interviewed the Italians, Pash returned to London on June 10 to prepare the Alsos Mission for the move to France in the wake of the Normandy invasion. He left behind the CIC agent who had traveled with him to Rome, Perry Bailey, to maintain an Alsos presence until Pash's executive officer, Maj. Richard Ham, could arrive there to establish a permanent Alsos office. Joining Ham by mid-June would be John Johnson and Robert Furman, both sent by Groves to continue the investigation into "military intelligence reports, scientific personnel, research centers and other institutions" in Rome and the surrounding areas.[93]

Upon his return, Pash learned of a planned change in administrative structure that threatened to undo Groves's careful consolidation of power in the field of atomic intelligence. Pash was told by Brigadier General Betts (G-2, SHAEF) that SHAEF was creating an advisory committee, known as the Combined Intelligence Priorities Committee (CIPC). The CIPC would consist of the members of the British Joint Intelligence Priorities Committee and an equal number of American representatives, and would be responsible for the evaluation of requests for technical or scientific intelligence throughout the theater. The committee would then prioritize these requests based on the CIPC's perception of their importance. To Pash, this would add a layer of bureaucracy that could delay, or even prevent, Alsos from accomplishing its mission.

To mitigate the potential damage to Alsos's effectiveness, Pash dealt with the CIPC in two ways. First, he stacked the U.S. contingent committee with Alsos members. Bryan Conrad was asked by SHAEF to designate the two army representatives, and Pash convinced him to choose Tony Calvert and Pash. One of the navy members was Capt. H. T. "Packy" Schade, Alsos's own senior navy scientist, and Samuel Goudsmit was placed on the committee to represent the OSRD. Second, Pash moved to sever atomic intelligence from the purview of the CIPC. He went to Bedell Smith and convinced him, and Betts, that atomic matters should be handled outside the bureaucracy of the CIPC. Instead, any action that required permission or logistical support would be submitted directly to SHAEF "for consideration and necessary action."[94]

Throughout the remainder of June and into July, Alsos continued to prepare for its move to the Continent, securing logistical support from theater combat units and supplementing its personnel with additional scientists and military officers. Major Ham arrived in London and was quickly dispatched to Rome to establish the mission's Mediterranean Section. Scientific Chief Samuel Goudsmit had arrived in early June, and his unit comprised scientists from the navy (Captain Schade, Capt. Wendell Roop, and Cdr. Jacob DenHartog), the War Department's New Developments Division (Mark May and Hans Reese), and the Army Service Forces (Col. Martin Chittick, Thomas Sherwood, Lt. Col. Edwin Foran, Lt. Col. Richard Ranger, and Capt. William Cromartie).[95] Pash's former executive officer in the Western Defense Command, Lt. Col. George Eckman, had been assigned to the mission and would serve in London as the deputy chief of mission. Eckman had served with Pash long enough that he could be depended on to represent Pash and Alsos with the CIPC and the British and U.S. military hierarchies. Rounding out the administrative/military group was Capt. Robert Blake (a tested combat veteran), Lt. Reginald Augustine (who had background knowledge of Western Europe and was fluent in several European languages), and Tony Calvert, whom Pash included in the operational group because,

Pash said, he had "made himself so valuable in developing the needed intelligence."[96]

Alsos France

The Alsos Mission in France began in earnest when Pash received an urgent message from Washington on August 5 reporting that Frédéric Joliot-Curie, referred to as "J" in Pash's official reports,[97] was thought to be at L'Arcouest in the Paimpol Area on the Brest Peninsula. On August 9, Pash and the CIC agent Gerry Beatson flew to Normandy and spent most of the next two days trying to get into L'Arcouest with elements of the Eighth Army Corps. Pash and Beatson joined with Task Force "A," the unit assigned to reduce German resistance at Paimpol and in the vicinity, and entered L'Arcouest on the morning of August 11. There, Pash reported, they discovered Joliot-Curie's house "to be totally cleared of all furniture and personal effects and the structure itself left in a very poor and dirty condition." It was clear that the most important French atomic physicist had not been there in quite some time.[98]

The following day Pash and Beatson moved on to Rennes in order to establish a base of operations there for the scientific group to wait until the liberation of Paris. There they set up billets and rations for the arriving Alsos scientists. While in Rennes, Alsos investigated the University of Rennes and, according to Groves, "discovered a number of catalogues and other papers that provided information pointing to possible future targets," including the town of Strasbourg in the Alsace region.[99] Tony Calvert and the Alsos interpreter PFC Nathaniel "Nat" Leonard systematically inspected the various offices and laboratories of the university. They discovered manuscripts, catalogues, and other literature relating to the science departments of several universities under German control, including German publications dated as late as March 1944. According to Pash, the documents captured "were to prove of considerable value, particularly in developing locations of German scientists and areas in which certain types of scientific research were being conducted."[100]

On what was most likely August 23,[101] Pash left Rennes with an advance party of Alsos, consisting of Calvert, Beatson, and Leonard, to link up with the U.S. Thirty-Eighth Cavalry Troop, whose assignment was to break through to Paris. On August 24 they found the Thirty-Eighth Cavalry, but quickly realized they would not be the first to reach Paris. Alsos pushed on to Longjumeau, where they joined the French Second Armored Division for the final push into Paris. Just before nine in the morning on August 25, Pash's team entered Paris behind three French tanks. The first American unit, and only the fourth Allied vehicle, to enter Paris since the fall of France in 1940 was a jeep with Pash, Calvert, and the two enlisted CIC agents of the Alsos Mission.

That afternoon Alsos moved to secure its primary target in Paris, Joliot-Curie. It was assumed that he would most likely be located in his laboratory in the College de France, on the Rue des Ecoles in central Paris. Pash made two attempts to break through to the university, but both failed due to heavy sniper fire and some remaining German resistance. Pash then retreated to French Army headquarters, where he tried to borrow French armored vehicles to push through the snipers, but his request was refused. Deciding that any further delay in securing Joliot-Curie was unacceptable, Pash and his team braved the snipers and by five in the afternoon they made it through to their objective. Frédéric Joliot-Curie was in American hands.[102]

Joliot-Curie and the scientists in his laboratory had been making homemade explosives (Molotov cocktails) for use by the French Resistance. He willingly spoke to the Americans and expressed his belief that the Germans had made little progress on uranium and were not close to making an atomic bomb,[103] although Goudsmit would later write that "it was plain that he knew nothing of what was going on in Germany."[104] He did provide Alsos with confirmation of intelligence it had collected previously, including the fact that two German scientists, Erich Schumann and Kurt Diebner, had visited Joliot-Curie and wanted to move his cyclotron and other scientific equipment back to Germany. Instead the two Germans kept the equipment in place and relocated to Paris to continue their research. A number of other prominent German physicists had also come to Paris, including Walther Bothe, a nuclear physicist at the Kaiser Wilhelm Institute for Medical Research; Abraham Essau, the head of physics at the German Ministry of Education in the Reich Research Council; Erich Bagge, a specialist in isotope separation; the experimental atomic physicist Werner Maurer; and Wolfgang Gentner, an authority on cyclotron operations who had worked with its inventor, Ernest Lawrence, in the United

States.[105] While this was not positive information about the progress of the German program, it was positive information about the existence of a program, or the Germans, Groves observed, "would not have found it expedient to use Joliot's laboratories."[106]

By the end of August, the full contingent of Alsos scientists had reached Paris, and Pash opened the Paris office of the Alsos Mission on the twenty-seventh. By this time it had become apparent that the current Alsos personnel did not have the physical capability to exploit all of the newly available intelligence targets, and so Pash requested additional assets. Through the assistance of Vannevar Bush and the OSRD, these personnel additions were approved, and by August 31, the Alsos Mission had grown to seven operations officers and thirty-three scientists, both civilian and military.[107]

In early September, Pash left the scientific team in Paris as the British and Canadian Armies were driving the Germans out of Belgium. Because this would make available certain key intelligence targets in Belgium, in particular the offices of Union Miniére Du Haut-Katanga, the Belgian mining firm that had shipped hundreds of tons of uranium to Belgium from the Congo, Pash felt it was necessary "to send a detachment of the Mission to Brussels and Antwerp to secure and consolidate the most important targets and to make arrangements for scientific personnel to exploit these targets."[108] Their mission, according to Pash, would be simple: "Get to Belgium without delay, determine where any stocks of refined uranium ore are located and in what amount, and seize any available supplies."[109] On September 5, a small unit consisting of Pash, Lieutenant Augustine, CIC special agents Carl Fiebig and Beatson, and interpreter Nat Leonard began their move from Paris to Belgium. Because their targets were in British occupied territory, and since contact with the British with regard to the activities of the mission had not yet been established, Colonel Conrad accompanied them. Pash and his group linked up with a British task force in Brussels and reported to its commander, a Colonel Strangeways. Through Conrad, whose "assistance," Pash reported, "was extremely beneficial and helpful and was responsible for the immediate cooperation" of Strangeways, Alsos was given permission to secure the offices and records of Union Miniére.[110] Gaston Andre, the head of uranium at Union Miniére's main office, gave Alsos valuable information about the movement of Belgian uranium to Germany during the war. He told Alsos that before the war, the Germans had bought less than a ton of refined uranium each month, but since June 1940, Groves later recalled, "orders from a number of German companies had increased spectacularly."[111] In all, more than a thousand tons of refined uranium had been shipped from Union Miniére to Germany since the start of the occupation.[112]

The ore shipped to Germany was, at least for the moment, beyond the reach of the Alsos Mission. Yet from Andre they had also learned that close to 150 tons of uranium had last been reported at Union Miniére's plant in Oolen, Belgium, and "might now be in the process of evacuation ahead of the oncoming Allied war machine," according to Pash.[113] On September 8, Pash reported these findings to Washington, and in response Groves immediately dispatched Furman to Brussels "with instructions to secure and ship the critical material."[114] He arrived in Brussels on September 17,[115] and the following day Furman and Pash met with Bedell Smith at SHAEF headquarters to plan the recovery mission. Also at the meeting was Maj. Gen. Kenneth Strong of the British Army, who was the G-2 for SHAEF. Strong sent a message to the G-2 of the British Twenty-First Army Group (the unit currently operating in the Brussels/Antwerp/Oolen vicinity) informing him that Pash "was on a mission of vital importance and that all necessary facilities were to be made available to him." According to Pash, this would allow him to operate in the British sector "without having to explain to any other person on the [British staff], or in the field, the nature of the Mission."[116]

From September 19 to September 25, Pash and his Alsos team searched Oolen for the Belgian ore, sometimes coming within two hundred yards of the German front lines while dodging sniper and mortar fire. On September 25 they located the sixty-eight tons of what Pash called the "desired material" and arranged for it to be shipped to the United States. They also learned that more than eighty tons of refined uranium had been shipped to France just prior to the German invasion. From September 26 to October 5, a force consisting of Pash, Furman, Major Vance, and Augustine (now a captain) searched the southwestern French countryside for the missing ore.[117] In Pash's memoir, he explained the mission's difficulty:

> My coded report to Washington provoked repeated exhortations to locate the lost eighty tons of uranium. But searching for freight cars that had been somewhere in France four years before was not that easy, especially since half of France was not yet under complete Allied control.

> So Alsos at the moment was supposed to mount an operation to secretly remove the uranium stocks from Belgium, to hit our Eindhoven targets when that city should fall into Allied hands, and to undertake a thorough reconnaissance of northern and southern France, the latter still under confused German and Free French control, in search of elusive freight cars. In addition, we had to supply adequate support for our scientific group at all times. The combined tasks seemed insurmountable.[118]

Despite these long odds, Pash and Alsos were able to find half of the material in Toulouse, France. Pash and Furman returned to Paris, but Major Vance, Captain Augustine, and Special Agent Fiebig remained in southern France to search for the remaining uranium. Although Alsos continued to receive information that the ore was in the area, however, it would never be located.[119] Pash and Alsos would later discover that the recovered uranium ore from Oolen and Toulouse would be used by the Manhattan Project to construct Little Boy, the atomic bomb dropped on Hiroshima on August 6, 1945.[120]

Throughout the remainder of October and into the beginning of November, Alsos continued to exploit scientific intelligence sources in France, Belgium, and (briefly) Holland. While doing so, Pash and Goudsmit planned for future operations in areas yet to be captured by Allied forces. Since Alsos had first entered France, one city on the border with Germany had been of particular interest to the mission's members: Strasbourg. In August, when Pash arrived on the Continent, he heard bits and pieces of information at the University of Rennes about a new institution built by the Germans in Strasbourg. This was enough to make it a future target of Alsos, but in Paris the mission was provided by the OSS with a university catalogue that indicated that three prominent German atomic scientists were working and teaching there: Rudolph Fleischmann, Werner Maurer, and, most importantly, Carl Friedrich von Weizsäcker. Pash wrote that Alsos scientists studied the catalogue as though it were "a spicy French novel—with photographs."[121] According to Goudsmit, "Everything seemed to point to the fact that the Germans were trying to transform the French institution into a model German university and that Strasbourg was a key target for us."[122] In late September, during a quick incursion into Holland during the failed Operation Market Garden, Alsos interviewed Dutch scientists at the Philips Laboratory in Eindhoven who told them that atomic research was, in fact, being conducted at the University of Strasbourg. In addition, Alsos discovered that special equipment for atomic research had been built at Philips and had been shipped to the university.[123]

As the U.S. Sixth Army Group moved eastward in November, it became clear that Strasbourg would soon be in the hands of the Allies. The Alsos Mission might at long last have the opportunity to discover some concrete evidence about the progress of the German atomic bomb program. Everyone involved with Alsos—from Vannevar Bush, Leslie Groves, and John Lansdale in Washington to Pash, Furman, Goudsmit, and the junior members of the team in the field—eagerly awaited the advance of the Allied armies. On November 25, 1944, the German Army abandoned Strasbourg, and the men of MED intelligence, led as always by Boris Pash, entered the city that would finally hold the key to unlocking the mysteries of Germany's atomic bomb.

4

TRANSITIONS

From the German Threat to the Soviet Menace

On the morning of November 25, 1944, the leading elements of the Alsos Mission—Boris Pash and the CIC agents Carl Fiebig and Gerry Beatson—entered the ancient city of Strasbourg. They immediately moved to secure the University of Strasbourg offices and laboratories. Once these were under control, Pash and his team left the university facilities under the guard of U.S. Army personnel and set out to track down the German scientists in their homes. Their first stop was the home of the German physicist Rudolph Fleischmann, where they were told by a neighbor that their quarry had left the city the day before. Alsos had equally poor luck at Carl von Weizsäcker's house, as well as the homes of the others on the Alsos target list. It appeared as though the German scientists had fled the city to escape the approaching Allied armies. And it seemed the long-awaited mission to Strasbourg, a city where Groves, Bush, and the leadership of the Manhattan Project intelligence team had placed such high hopes, would end in failure.[1]

The scientific group, headed by Samuel Goudsmit, had remained behind in Paris to await word that Pash had captured the German scientists.

Instead Goudsmit learned that the Germans were gone. He reluctantly passed this unfortunate news to Vannevar Bush, who was then in Europe for a brief visit. The message of the failure at Strasbourg would reach Leslie Groves through Robert Furman three days later: "Pash wired from Strasbourg that friends there have departed."[2]

As it happened, by the time Groves received this news, the situation in Strasbourg had significantly changed. While Pash was disappointed by his initial failure to capture the German atomic scientists, he refused to accept defeat. On the same day that he had unsuccessfully searched Fleischmann's home for the prominent German physicist, Pash learned that another German atomic scientist was in the city.[3] Although not particularly well known or prestigious, this man was still on Pash's master list of German scientists of interest, and the Alsos team immediately hunted him down. Nervous and evasive, the German refused to give Alsos any information. Yet just as Pash was ready to wrap up the questioning, the scientist asked if he could go to the Strasbourg Hospital the following day. He explained that some of his laboratory work was performed at the hospital, and Pash's intelligence instincts, honed by years of work in counterintelligence and Alsos, convinced him that the hospital could be the key to turning a failed operation into an overwhelming success.

The following morning, November 26, Pash and his team stormed into the hospital director's office and demanded to see Fleischmann's atomic laboratory. This was a bluff: Pash had no real concrete information that Fleischmann or anyone else was at the hospital conducting atomic research. But the bold move paid off. The cowed hospital director led Pash to a separate wing of the grounds, where he discovered Fleischmann and five other German scientists on his list. The Germans had been hiding in the hospital and were wearing medical clothing to pass themselves off as hospital staff. Of the atomic scientists, only Weizsäcker was missing; he had indeed left before the city was captured by the Allies.[4]

In response to this news, Goudsmit and the DuPont chemist Fred Wardenberg (an Alsos scientific member) immediately set out for Strasbourg. Delayed by fierce German resistance in the area, which at one point threatened to push the Allies out of Strasbourg, Goudsmit arrived on December 3 and found that Pash and Robert Furman (who had experienced less difficulty than the scientists negotiating German fire and arrived days earlier) had already collected thousands of German scientific documents.

Together, Goudsmit, Wardenberg, and Furman began a systematic interrogation of the German scientists and some of their support personnel, such as Weizsäcker's secretary.

In the end, these interviews provided very little in the way of actionable intelligence. Fleischmann and the German scientists refused to give Alsos any clues to either the location of the German atomic research center or its status, and the support personnel, while more open with their interrogators, were ignorant of the activities of the German atomic physicists. Fortunately for Alsos, the captured documents and personal correspondence Pash found in offices, laboratories, and homes provided the mission with the intelligence breakthrough they had hoped Strasbourg could deliver. For one thing, the documents revealed the locations of the remaining German atomic scientists on Alsos's target list. One piece of paper had the letterhead of the Kaiser Wilhelm Institute for Physics (Werner Heisenberg's institute), and it showed that the key target had been evacuated from Berlin to the small village of Hechingen in Württemberg. The paper even provided the precise address and telephone number for the secret German laboratory. Other documents hinted that Weizsäcker and Karl Wirtz, an expert on heavy water and isotope separation, had joined Heisenberg in Hechingen, and that Otto Hahn's laboratory had been moved to the town of Tailfingen. Letters from scientists originating in the German communities of Stadtilm in Thuringia and Bisingen in Württemberg, along with references to secret caves in Haigerloch, gave Alsos a number of future targets.[5]

Far more important than even the location of Heisenberg and the other German atomic scientists was the intelligence gained through the Strasbourg documents regarding the status of the German atomic bomb program. According to Pash, this was "probably the most significant single piece of military intelligence developed throughout the war."[6] Through documents and correspondence the Alsos team was able to discern that Germany was having major difficulties with the separation of uranium isotopes, and had yet to separate U-235 in any amount even remotely significant for bomb manufacture. In fact, the documents showed that as late as August 1944, the Germans had only just recently begun their atomic pile (reactor) work, and were still some ways away from achieving a self-sustaining chain reaction, a benchmark the U.S. project had achieved almost two years earlier. The Germans were still unsure of the correct reactor design, and their early experiments had not given them hints to the problems that Enrico Fermi's team in Chicago had had to overcome before they could get the U.S. reactor online. According to Goudsmit, "In short, they were about as far as we were in 1940, before we had begun any large-scale efforts on the atomic bomb at all." Despite the fact that the documents also showed that the German leadership had given the project a high priority, and that the German Army was taking part in the research, "as far as the German scientists were concerned, the whole thing was still on an academic scale."[7]

For the Alsos Mission's scientific chief, "the conclusions were unmistakable." The documents and letters captured at Strasbourg "proved definitely that Germany had no atomic bomb" and would not be able to produce one before the end of the war. They were not even far enough along in their research to present a danger from radiological attack.[8] Goudsmit would later joke about the dismal state of the German program, "Sometimes we wondered if our government had not spent more money on our intelligence mission than the Germans had spent on their whole project."[9]

Boris Pash was convinced as well. He would write in his memoirs that at Strasbourg, "Alsos exploded the Nazi super-weapon myth that had so alarmed Allied leaders. . . . Alone, that information was enough to fully justify Alsos."[10] He and Goudsmit decided to deliver the Strasbourg results in person to the Allied headquarters in Vittel, France, where Vannevar Bush had come up from Paris to get the results of the mission, and where Pash could use the secure communications system to send the report to General Eisenhower and General Groves. As they made the drive to Vittel, Pash could not help thinking, "*Alsos has exploded the biggest intelligence bombshell of the war!* Now every American and British leader in the know would sleep more comfortably."[11]

Goudsmit explained to Bush what the Strasbourg documents revealed, and after Goudsmit effectively answered some of Bush's questions, Bush too was a believer. In Paris he met with Bedell Smith, who outlined Eisenhower's plan for the remainder of the war. Smith asked Bush if he needed to press Eisenhower to speed up the attack plan, risking heavier casualties but ensuring the war's end before the Germans could use an atomic bomb. Bush, confident in the Strasbourg results, explained to Smith that the Americans were well ahead of the Germans in the atomic field. "In fact," Bush would explain in his memoirs, "we were so far ahead that their effort, by comparison, was pitiful."[12] He would later estimate that the Germans had only achieved "five percent" of what the Americans had accomplished in atomic research and development.[13] Bush told Smith that Eisenhower, if he so desired, "could take a couple more years, if necessary" to win the war. "There would be no German atomic bomb."[14]

Leslie Groves, who directed all U.S. atomic intelligence, would be the final arbiter of the Strasbourg intelligence. While he did not get to see the raw data provided in the documents and correspondence, Groves had come to trust Pash, Goudsmit, and certainly Bush. Based on their recommendation, he reported the Strasbourg findings to General Bissell, explaining that the intelligence was "the most complete dependable and factual information we have obtained bearing upon the nature and extent of the German effort in our field." Groves continued, "Fortunately, it tends to confirm our conclusion that the Germans are now behind us."[15] While this language was somewhat guarded, in his memoirs Groves was more steadfast about his beliefs. There he stated that "all evidence from Strasbourg clearly pointed to the fact that, as of the latter part of 1944, the enemy's efforts to develop a bomb were still in the experimental stages, and greatly increased our belief that there was little probability of any sudden nuclear surprise from Germany."[16]

Zurich

If any doubt remained that the Germans were not close to building an atomic bomb, that question was put to rest by Moe Berg in December 1944. Either in late November or early December (the exact date is unclear), Allen Dulles in Bern learned that Werner Heisenberg, assumed to be the leader of the German atomic bomb project, would be in Zurich on December 18 to give a lecture to professors and graduate students at the university. The invitation for the lecture was issued by a friend of the OSS, Paul Scherrer, who was also most likely the origin of the intelligence on Heisenberg. Regardless of the exact details, Groves and Furman saw an opportunity to either confirm the Strasbourg results or, if Heisenberg and the Germans were actually in the process of creating an atomic bomb, permanently remove Heisenberg from his leadership position in German atomic physics.

Berg arrived at the lecture hall at the Federal Technical College on the day of Heisenberg's visit. In the guise of a physics graduate student, he found a seat in the audience behind Otto Hahn and Carl von Weizsäcker, who had accompanied Heisenberg to Zurich. Berg carried a pistol in his pocket; his mission, as was briefed to him by Robert Furman before his departure, was to kill Heisenberg if he became convinced the Germans were close to building an atomic bomb. Killing Hahn and Weizsäcker would be an added bonus, unanticipated by MED intelligence, but certainly an act that would be welcomed by Furman and Groves.

Heisenberg lectured on the advanced physics principle of S-matrix theory, a topic far removed from atomic bomb manufacture, and covered matters that exceeded Berg's basic physics knowledge. Nonetheless, Berg's cover held long enough for him to be invited to Scherrer's dinner party for Heisenberg, which followed the lecture. At the party, Berg heard Heisenberg announce that he believed the war was all but over, that Germany would almost certainly lose, and that his primary focus was rebuilding German science after the war concluded. Later in the evening, Berg arranged it so that both he and Heisenberg left the party at the same time. They walked back to their quarters and discussed Heisenberg's latest physics research (Berg's linguistic skills were apparently so refined that Heisenberg did not seem to detect any trace of an American accent). At the end of their walk, Berg was convinced that the German atomic bomb program did not exist in any way that could threaten the Allies during the war. Based on this conclusion, and Heisenberg's declaration of the expected German defeat, Berg decided not to assassinate Heisenberg.[17]

When Groves was placed in charge of gathering all foreign atomic intelligence in September 1943, his primary task was to ascertain the progress made by Germany in the research and production of atomic weapons. The Alsos Mission, which was launched in December, was the culmination of this mandate: its sole purpose was to discover the extent of German progress toward the development of an atomic bomb. By the end of 1944, Alsos had successfully accomplished this objective. Why, then, wasn't the Alsos Mission disbanded following the completion of the Strasbourg operation, and its members reassigned to units or scientific projects where they could have a positive impact on the effort to end the war? Robert Furman assumed this would be the case. In the autumn of 1944, just prior to the Strasbourg mission, the officer Groves trusted most to run the day-to-day operations of his foreign intelligence organization wrote Groves that "as soon as it is indicated that there has been no German progress in the project field, I will, at the proper time, close the mission down as far as the project is concerned and return our personnel to the United States."[18] Samuel Goudsmit had also been given the impression that the mission would end when the United States was convinced the Germans had no bomb. In a letter to his wife and daughter dated December 10, 1944, Goudsmit hinted that the "unsuspected success" of the Strasbourg operation would mean a "short trip home." He was so sure Alsos would be terminated that he speculated that he might make it back to the United States "just around X-mas," maybe even at the same time the letter reached his family.[19]

But the Alsos Mission would not be disbanded. In fact, the U.S. government decided to *increase* the manpower and resource allocation for Alsos and to formalize its organizational structure in the first months of 1945. In January, the mission was given, for the first time, its own table of organization and equipment with, Pash wrote, "a definite allowance of personnel and specific items of equipment." The mission would no longer have to borrow equipment from U.S. combat units in order to continue to function.[20] It was also allotted additional administrative personnel, most importantly a deputy mission chief and a deputy scientific chief who would serve as administrative lieutenants to Pash and Goudsmit. These deputies would, according to the minutes of a committee meeting in December 1944, "staff the Mission headquarters [in Paris] at all times," allowing Pash and Goudsmit to continue leading field operations "while effective administrative control over the Mission's operations may be maintained."[21]

By February, the Alsos Mission had established two forward bases that supplemented the main mission office in Paris. Alsos Forward South in Strasbourg was commanded by Capt. Reginald "Reg" Augustine, and Alsos Forward North was established in Aachen under the command of Maj. Russ Fisher. The plan was to shift the forward bases farther into Germany in accordance with the movements of the Allied combat units in order to, also based on the December 1944 meeting minutes, "approximately conform to the concentration of German scientific and industrial centers."[22] Pash decided to keep the main Alsos office in Paris for several reasons. First, since Alsos would have to operate within the zones of three different army groups, each of whose activities were coordinated and commanded by SHAEF (whose headquarters were located outside Paris at Versailles), it made sense for the Alsos Mission to maintain continual contact with Eisenhower's office. In addition, most of the other Allied organizations essential to Alsos's success had their theater headquarters in Paris, including the Office of Scientific Research and Development, the Navy Technical Mission, and the Office of Strategic Services.[23] Finally, the Paris office served as a communications hub for Alsos. Messages from the Pentagon (in most cases from Groves) came to the theater through the Paris headquarters of the European Theater of Operations, U.S. Army, and the Alsos office, Pash observed, allowed the mission "to maintain the initiative in dealing with Washington."[24]

The reason the U.S intelligence leadership had decided to keep the Alsos Mission in the field was that U.S. atomic intelligence had shifted its primary focus. Since Pash and Goudsmit had discovered the insignificance of the German atomic bomb program at Strasbourg, the Alsos Mission had shifted its operational attention to preventing the Soviet Union from acquiring the means to build its own atomic weapons. To be sure, there were some in Washington who wished to capture all the remaining German uranium, laboratories, and atomic scientists in an effort to remove, with absolute certainty, any doubt about the German atomic program. Yet in the last months of the European war, Alsos would engage in operations designed to deny the Soviet Union the knowledge, men, and materials necessary to become an atomic power—namely, by capturing German uranium ore, locating secret laboratories, and interning prominent German atomic scientists so that, according to Pash, "they should not fall into Soviet hands!"[25]

The Soviet Union's Interest in Atomic Weapons

Much like their American, German, British, and French counterparts, Soviet scientists immediately understood the implications of the discovery of nuclear fission. In 1939 they began the task of replicating Otto Hahn and Fritz Strassmann's fission experiment, and worked through the calculations surrounding uranium enrichment and the possibilities of self-sustaining chain reactions. By the summer of 1941, however, the German invasion of the Soviet Union forced the nascent Soviet atomic research program to a premature halt. Physicists who had been working on the nuclear problem were reassigned to more practical (and immediate) war work. The Soviet Union did not have the resources to launch a Manhattan Project–scale program while simultaneously fending off a German invasion. The little work that was done on the nuclear question was confined to sporadic laboratory-scale experimentation by scientists who could be spared from the war effort. The Soviet intelligence services, however, did not have the same restrictions, and they set their sights on the United States and on stealing the secrets of the atomic bomb.[26]

Early in the war, the U.S. government learned of Soviet interest in the U.S. atomic bomb program. American officials understood that, according to a MED counterintelligence summary, the Soviet Union "through its Embassy officials and espionage agents in the United States has been active for a long time trying to elicit as much information as possible concerning the project."[27] When Leslie Groves was given command of the Manhattan Project in the summer of 1942, he was told that the Soviet Union had an ongoing operation to discover the secrets of U.S. fission research. He was instructed to maintain a strenuous counterintelligence program, designed, of course, to keep the Germans in the dark about the project. Yet an equally important task, he later wrote, was "to keep the Russians from learning of our discoveries and the details of our designs and processes."[28]

The majority of information about the Soviet espionage efforts came from none other than Lt. Col. Boris Pash. In his counterintelligence work as an officer in the Western Defense Command, Pash established an elaborate system of investigation into Communist espionage activities surrounding the U.S. bomb program. According to John Lansdale, by early 1943 Pash was "conducting a wide spread and complex investigation of communist activity" primarily located at the Radiation Laboratory of the University of California, including "intense investigations of the activities of several Communist Party members" who were working on the Manhattan Project. Pash and his team followed the suspected Soviet agents, and installed microphones in their houses and the places they frequented. According to Lansdale, Pash and his team installed "a new cord on the telephone receiver with one more wire than ordinarily required. This wire bypassed the disconnect on the telephone enabling the telephone to be used as a microphone." This allowed U.S. counterintelligence to monitor Soviet telephone calls, as well as conversations that took place near the telephone.[29]

Through these methods, Pash was able to discover that several members of the Berkeley Radiation Laboratory had passed secret information to Steve Nelson, a member of the National Committee of the Communist Party USA and leader of the Communist Party in Alameda, California. Nelson, who had been the political commissar of the Abraham Lincoln Brigade of the Republican Army in the Spanish Civil War, spent the first years of the Second World War directing the efforts of the Federation of Architects, Engineers, Chemists, and Technicians, a Communist front organization that according to Groves was "making extraordinary efforts" to organize the laboratory at Berkeley. These activities, Groves noted, were designed "to the extent of securing and training perspective employees" for espionage work.[30] Nelson had been observed on several occasions meeting with contacts at the Soviet Consulate in San Francisco and the Soviet Embassy in Washington, presumably to pass along information received from project scientists.[31]

Throughout the war, suspected Soviet espionage was investigated across the United States, from the laboratories in California to the Metallurgical Laboratory at the University of Chicago to scientific research centers at Columbia University in New York. In each case, Communist scientists (or at least scientists with Communist sympathies) were observed meeting with members of the Soviet diplomatic delegation or with the leadership of the Communist Party USA. Along with the MED counterintelligence team, the Federal Bureau of Investigation tracked the Soviet operation nationwide. According to FBI director J. Edgar Hoover, "During the period that the Army has been engaged in the supervision of this experimentation, numerous efforts have been made by the Soviets to obtain the highly secret information concerning the experimentation and this Bureau has been actively following such Soviet efforts."[32]

The Soviet Union also demonstrated its interest in atomic weapons through what Groves described as "unsuccessful attempts to secure uranium concentrates in [the United States] both through private firms and officially through the Lend-Lease Administration."[33] Although the Soviet

Purchasing Commission alleged that the uranium was for uses not related to uranium fission, Groves was suspicious enough to initially deny its January 1943 request for twenty-five pounds. After the Soviets complained to the Lend-Lease Administration in March, Groves grudgingly acceded to their demand, agreeing to send one kilogram (2.2 pounds) of lower-quality uranium metal. Even then, Groves did not deliver the uranium until February 1945, more than two years after it was originally requested.[34]

Completing the intelligence picture on Soviet intentions were American scientists who had traveled to the Soviet Union during the war and had returned to report their interactions with their Soviet counterparts. According to Groves, Soviet scientists were "unduly curious in their questioning of American scientists visiting Russia concerning our work on uranium fission." In addition, American scientists told the MED intelligence team that the Soviets had constructed their own cyclotron, an indication of at least a nascent atomic weapons program.[35]

Taken together, Soviet espionage attempts, combined with Soviet interest in fissionable materials acquisition, the construction of a cyclotron, and the Soviets' interrogation of American scientists, presented Groves and MED intelligence with a considerable counterintelligence problem throughout the war. The Germans were, of course, the primary focus of counterintelligence efforts. Yet as John Lansdale explained, "From the beginning, Russia was regarded, from an intelligence standpoint, as an enemy."[36] Because the Germans had been eliminated as an atomic threat after the Alsos Mission's discoveries at Strasbourg, Groves, Lansdale, Furman, and Pash could concentrate their full efforts on the Soviet Union.

But they still had pressing issues in Western Europe. In 1940, when the fall of France to the German Army was imminent, Frédéric Joliot-Curie sent several of his fellow French atomic scientists to Great Britain to prevent them from being captured by the Nazis. He wanted to prevent, as much as he could, French atomic secrets and key atomic materials from falling into the hands of the Germans. Along with the scientists, Joliot-Curie sent France's entire stock of heavy water (at the time the largest in the world), all of their research reports, and two grams of radium to Britain. Some of these French scientists continued on to work for the British atomic bomb program in Montreal, Canada, which worked in coordination with the Manhattan Project. Others came directly to the United States, where they worked with American scientists to further Allied research into nuclear development. All of these men remained in close contact with one another during their time of exile in Canada and the United States, and all indicated their intention to return to France once it had been liberated from German occupation.

This repatriation would begin shortly after the Allies established a firm foothold in France in the late summer of 1944. The first to return was Pierre Auger, a French physicist who had worked on the atomic reactor project in Montreal. He was followed shortly thereafter by Jules Gueron in October. Gueron had worked with American scientists and had learned a great deal about the U.S. atomic bomb program. In November, the French scientist Hans von Halban, also a member of the British portion of the Allied atomic bomb project, requested a return to France to see and report to Joliot-Curie. The British agreed, and Halban met with Joliot-Curie and, according to Groves, disclosed to him "vital information concerning the Project," including information concerning "data and research that had been developed by American funds and effort."[37]

The issue at hand was Frédéric Joliot-Curie. After the fall of France, he had been actively involved in the French Resistance movement and had assisted the underground by developing technical methods for sabotage and communication. Once France was liberated, Joliot-Curie, the preeminent scientist in all of France, was appointed director of scientific research in the Provisional Government of the French Republic.

He was also a Communist. As a supporter of the Resistance, Joliot-Curie joined the Communist Party in the spring of 1944 and publicly announced his membership in the party in August. According to Alsos Mission notes taken during the interview of Joliot-Curie in September following the liberation of Paris, the French physicist had "very strong political views and frankly declared that he is a 'communist.'"[38] As a prominent figure in the newly liberated France, Joliot-Curie was even elected a member of the French Chamber of Deputies—as a member of the French Communist Party.[39] Complicating matters even further was the fact that, according to one intelligence report, Frédéric's wife, Irene Joliot-Curie, was considered "undoubtedly more dynamic politically than her husband." In a word, U.S. intelligence understood Irene to be a "fanatic" who had been a Communist for years, surrounded herself with "scientists of the extreme-left," and had used her influence over her husband to push him into the Communist Party.[40]

Taken together, the Joliot-Curies presented a real problem for Leslie Groves and U.S. atomic intelligence. Their political affiliation forced Groves to assume that any secret atomic information learned by the French would immediately be passed along to the Soviet Union.[41] This meant not only that Groves would have to pressure the British to prevent any further exchange of information between French expatriate scientists and the Joliot-Curies, but that he would also have to pay close attention to the progress of the French military forces as they began to push into Germany. If the French Army was allowed to capture German atomic facilities, materials, or scientific personnel, it could detrimentally affect U.S. security as much as if the Soviet Union directly assumed control of those laboratories, uranium supplies, or atomic physicists.

Fortunately, Groves had an organization already in place in Europe that was uniquely trained and equipped to secure outstanding German nuclear resources: the Alsos Mission. Under the leadership of Boris Pash and Sam Goudsmit, it would spend the better part of 1945 denying the Soviets, and the French, the fruits of German scientific expertise.

In February 1945, the Allied armies began to make their long-awaited push into Germany. The Americans, British, and French entered from the West and advanced toward the Rhine, while the Soviets invaded Pomerania and Silesia in the East. By March, the Allied forces in the West had crossed the Rhine and were pushing toward Alsos's first major German target, the university city of Heidelberg, which housed the Kaiser Wilhelm Institute for Medical Research. As Alsos waited for Heidelberg to fall to the Allies, Groves in Washington was devising a plan to deny the Soviet Union a key German atomic resource.

At the Yalta Conference of February 4, 1945, President Roosevelt, Prime Minister Churchill, and Premier Stalin agreed on the postwar occupation and partition of Germany. Although most of the Alsos Mission's objectives were located in the American, British, or French zones of advance, the town of Oranienburg, located about fifteen miles north of Berlin, would be in the Soviet zone of occupation, and this presented a real problem for Groves and MED intelligence. Oranienburg was the home of the Auergesellschaft Works, a German industrial plant that was, Groves reported to George Marshall, "manufacturing by highly secret processes certain special metals to be used for the production of as yet unused secret weapons of untold potentialities"—Auergesellschaft produced uranium for bomb research.[42] The Alsos Mission, which could only advance as quickly as the Allied armies, would not be able to reach the factory before the Soviet Army. According to Groves, "There was not even the remotest possibility that Alsos could seize the work,"[43] and so he decided it would be necessary to destroy the plant before these important materials fell into Soviet hands.

On March 7, Groves sent one of his officers, Maj. Francis J. Smith, from Washington to London to explain the mission, in person, to Gen. Carl "Tooey" Spaatz of the U.S. Army Air Forces. Spaatz, commander of the Strategic Air Forces in Europe, was told in a memorandum from Marshall to expect Smith, who would advise him of the "reasons for bombing a certain vital target." Marshall warned Spaatz that the "matter is of the highest order of secrecy," and implored him to refrain from informing anyone, including his own officers, of the true purpose of the bombing.[44] Spaatz was convinced of the importance of the secret mission, and immediately created an operational plan for the destruction of Oranienburg.

According to Groves, on March 15, 1945, Spaatz dispatched 612 B-17 Flying Fortresses and B-24 Liberators of the Eighth Air Force to destroy the Auergesellschaft Works. Escorted by 782 fighter aircraft, the bombers dropped 1,684 tons of incendiary and high-explosive munitions of "varying sizes up to 2000 lbs" on Oranienburg "with a wide range of fuzings including long delays" so that dust and smoke would not "obscure the target for the formations which followed."[45] Four days later, on March 19, Spaatz wrote Marshall and informed him that "the results of the attack on the special target at Oranienburg are excellent." Poststrike reconnaissance of the target area flown the day after the attack showed "virtual destruction" of the Auergesellschaft Works. It appeared "that substantially all of the buildings within the special target area are gutted or burned out," and photographs showed that all parts of the plant located aboveground had been completely destroyed. "In general," Spaatz wrote, "it is fair to say we are extremely pleased with the indicated results" of the attack.[46] The Soviets would not get their factory.

Of course, the true purpose of the mission would be obvious to anyone paying attention. Oranienburg had no real strategic value to Allied war aims, and the Auergesellschaft Works was the only legitimate target in the city. Groves had always made a concerted effort to disguise U.S. interest in atomic research from the Germans, and now he was faced with having to make the same calculated decisions regarding the Soviet Union. Therefore, to conceal the real purpose of the Oranienburg mission from the Soviets, Groves recommended, Marshall approved, and Spaatz planned a simultaneous and equally heavy attack against the small German town of Zossen, home of the German Army headquarters. The attack, which consisted of 735 bombers, achieved its intended purposes, and according to Spaatz "drew most of the attention and in itself presented a plausible cover plan for the Oranienburg operation."[47] As an added, and unexpected, bonus, Groves would learn after the war that the Zossen raid severely wounded and incapacitated Gen. Heinz Guderian, chief of the German General Staff.[48]

At about the same time Spaatz was reporting the bombing assessment to Marshall, Allied forces captured Heidelberg. The Alsos Mission, riding on the heels of the advancing combat forces, immediately moved into the city to secure the laboratory of Walther Bothe, a prominent German nuclear experimental physicist. He and his laboratory were located in the Physics Department of the Kaiser Wilhelm Institute for Medical Research, and when Alsos arrived Bothe quietly accepted his capture. He willingly spoke to Sam Goudsmit about the German atomic program, told him about the research work done in his institute during the war, and showed him what Goudsmit later described as "reprints, proofs and manuscripts of all the war-time papers which were written under his direction."[49] Bothe also revealed, or at least confirmed, the location of the remaining German atomic scientists. Otto Hahn had been evacuated from Berlin to Tailfingen, a small town about forty miles south of Stuttgart, near Hechingen (all in southern Germany). The German experimental uranium reactor (atomic "pile") had been removed from the Berlin area and shipped to the town of Haigerloch, also in the vicinity of Hechingen. In Hechingen itself were Max von Laue and the crown jewel of Alsos targets, Werner Heisenberg. Clearly, the future operations of the Alsos Mission would center on southern Germany, in and around the city of Hechingen.[50]

Bothe also confirmed to Goudsmit the sorry state of the German atomic bomb program. In Bothe's laboratory was the only German cyclotron in operating condition (in contrast, the United States had twenty of these key machines for nuclear research).[51] He reported a shortage of heavy water, the only major source of which was destroyed by Groves when the United States bombed the Norsk Hydro plant in Norway.

Finally, Bothe told Goudsmit that the total German effort on atomic bomb research consisted of only a handful of scientists: his group in Heidelberg, Heisenberg in Hechingen with ten other subordinate physicists, a man named Dopel in Leipzig (who was assisted by his wife), a man named Kirchner in Germisch with two assistants, and a physicist named Stetter in Vienna with five others. Otto Hahn, according to Bothe, was working on chemical research not associated with the German nuclear program.[52]

On March 30, Pash moved the Alsos Forward South Base from Strasbourg to Heidelberg in order to bring the administrative structure of the mission closer to the front lines. Joining Pash in Heidelberg was Major Ham, who would assume control of administration and planning in the southern area so that Pash and Goudsmit could continue field operations.[53] Soon after the base was established, Pash learned that George Patton's Third Army was rapidly advancing through central Germany and that the city of Stadtilm, in Thuringia, would soon fall into the hands of the Allies. Alsos believed that the Germans had built a secret experimental atomic pile there, in a laboratory created by Army Ordnance. They also had information that led them to believe that the laboratory housed two prominent German atomic scientists, Kurt Diebner and Walther Gerlach (whom Goudsmit described as the "chief co-ordinator of nuclear research" in Germany).[54]

During the first week of April, Alsos moved into Stadtilm and located the Army Ordnance laboratory. There they discovered the German uranium pile in the cellar of an old schoolhouse. In the center of the cellar the Germans had dug a deep pit, and had planned to build a reactor of uranium oxide blocks surrounded by heavy water. According to Goudsmit, the whole operation was "on the scale of a rather poor university and not of a serious atomic energy project." While Alsos was able to capture several German physicists and their families, the mission's two primary targets, Diebner and Gerlach, were gone, along with most of their materials and equipment. The captured researchers told Goudsmit that Gerlach had been gone for some time, but that Alsos had just missed Diebner, who had left only two days earlier. Apparently the Gestapo, just prior to the fall of the city, had taken Diebner, his materials, and his research documents and relocated them to Bavaria, where the captured scientists assumed he would be asked to resume his research.[55]

Pash, Goudsmit, and the Alsos Mission had little time to dwell on their disappointment at missing out on the two German scientists. During the same time that Alsos was searching the Army Ordnance laboratory in Stadtilm, Leslie Groves was dealing with a major diplomatic problem in Washington. The Yalta Conference had divided Germany into three zones of occupation. Later it was decided to include a fourth zone for the French out of territory originally intended for the United States. This proposed French zone would include the four towns (Hechingen, Tailfingen, Haigerloch, and Bisingen) where it was believed the majority of the remainder of the German atomic bomb program was located, including almost all of the top German nuclear scientists. By late March, the French Army was poised to move into this area. To Groves, this would be a disaster. His knowledge of Frédéric Joliot-Curie's politics, he later wrote, had convinced him "that nothing that might be of interest to the Russians should ever be allowed to fall into French hands."[56]

After consulting with George Marshall, on April 3 Groves wrote a letter to Secretary Stimson, pleading with him to intervene with the State Department, which was responsible for the readjustment of the American zone's boundaries. Groves asked Stimson to convince the State Department to retain in the American zone the "quadrilateral of Freiberg, Stuttgart, Ulm, Friedrichshafen" (within which were the four target cities).[57] Stimson passed this request on to Secretary of State Edward Stettinius, arguing that it was of the "highest importance" that the United States keep this important territory.[58] However, despite protests from the secretary of war, the army chief of staff, and the head of the Manhattan Project, the State Department refused to consider moving the boundaries without a full explanation of the reasons why the request was being made, something Groves would never give.[59] At any rate, it was unlikely that the French would agree to any reshuffling of the assigned territory. They were, according to Lansdale, "extremely anxious" to move into the area, since the Vichy French government in exile was located in the vicinity of Lake Constance in the southern portion of the disputed territory.[60] Abandoning all hope of moving the zones of occupation, Groves was forced to initiate a dramatic measure to accomplish his purposes: Operation Harborage.[61]

Operation Harborage was designed to get Alsos members into the key target cities before the French forces so that they could, as Goudsmit explained, "capture the people [they] wanted, question them, seize and remove their records, and obliterate all remaining facilities."[62] The plan called for the Alsos Mission to be attached to a reinforced corps (two armored divisions, an airborne division, and all of the necessary logistical support) that would cut diagonally across the front of the French lines. Groves sent John Lansdale to Europe to make the necessary arrangements for Harborage. Lansdale left Washington on April 6 and arrived in Paris on the eighth. He immediately reported to Gen. Bedell Smith in Rheims and described to him the nature of the mission, explaining that the "apparent untrustworthiness and bad associations of many of the French personnel" made it imperative that U.S. forces capture the area before the French. Lansdale told Smith that the mission was "deemed by the War Department highly important," but that Eisenhower would have final discretion as to whether or not it could be executed without detrimentally affecting the overall strategic picture. Smith told Lansdale that Boris Pash had already briefed him on Harborage. Smith had already sent Pash to confer on the plan with the Sixth Army Headquarters.[63]

On April 10, Lansdale, Pash, and Furman returned to SHAEF headquarters and spoke to British Maj. Gen. Kenneth Strong, G-2 for SHAEF. They discussed the intelligence data, and Strong agreed that the information on the target cities was, Lansdale wrote, "definite and clearly indicated the presence of research activities in the area."[64] Lansdale and Pash were then brought into a staff planning meeting of top commanders to discuss Operation Harborage. Bedell Smith presided over the meeting, which included Strong; Gen. Harold Bull, operations officer (G-3) for SHAEF; General Craig, the head of the Operations Division, General Staff, War Department; and John McCloy, the assistant secretary of war. After each man had been given the opportunity to speak on the proposed operation, Smith stated that since the Sixth Army was at that moment required for defense purposes in the South (while the main U.S. thrust was in the North), he could not recommend the mission to Eisenhower at that time. He did say, however, that he would instruct his staff to construct the full operations plan for Harborage in case the strategic situation changed.[65]

Pressed by Lansdale, Smith also agreed to hold the Thirteenth Airborne Division in reserve to drop into the area in support of the French once the French Army began its advance (Bull promised that the Thirteenth could be ready to move within seventy-two hours). This could give Alsos the time necessary to move into the key cities and capture the target personnel, facilities, and materials. Finally, failing either of those options, Smith agreed to order a bombing mission against the targets so that nothing of value was left for the French. Lansdale's priority was to seize the German assets, but if that was impossible he felt it was absolutely imperative to ensure that they were "destroyed to the fullest extent."[66]

Four days later, on April 14, Pash, Furman, and Lansdale learned that these contingency plans would be unnecessary. The strategic situation in northern Germany had changed dramatically. Eisenhower decided to hold up the western Allied forces short of Berlin, and instead of continuing to press forward the armies would spend some time reinforcing their flanks. As a result, Eisenhower decided to put the original Harborage plan in operation. Once the French began moving again in the South, the assigned corps of Operation Harborage would sweep across their front and capture the target cities. Smith told Pash and Lansdale that he believed the operation could happen within as little as two weeks, although it could be longer.[67]

While Alsos waited for Operation Harborage to commence, Lansdale and Pash learned that an American force, the Eighty-Third Infantry Division, was closing in on the town of Stassfurt, in eastern Germany. Alsos investigations at Brussels in September 1944 had indicated that the supply of uranium captured by the Germans in Belgium (as much as 1,200 tons) had been sent to storage at the Wirtschaftliche Forschungs Gelleschaft salt mine in Stassfurt.[68] The city would soon become a part of the Russian zone of occupation, so it was essential that Alsos get there first.

Under the direction of Groves in Washington, Lansdale established an improvised joint U.S.-British task force to capture the Stassfurt uranium. The American members were Lansdale, Pash, Tony Calvert (whose intelligence had first discovered that the ore had been shipped to Stassfurt), several Alsos CIC agents (unnamed in the documents), and Maj. J. C. Bullock, whom Groves had transferred from the Manhattan Project to the Alsos Mission "for the express purpose," he said, "of recovering" the ore.[69] The British contingent included Sir Charles Hambro, a top adviser to the British government on raw materials (in particular uranium); Michael Perrin, the assistant director of British Tube Alloys (the British equivalent of the Manhattan Project); and Perrin's assistant, David Gattiker. The importance of the mission was demonstrated by the prestigious composition of the task force.[70]

On April 15, Lansdale, Pash, Bullock, and Hambro met with Brig. Gen. Edwin Sibert, G-2, Twelfth Army Group, to discuss the proposed operation (Stassfurt was in the Twelfth Army Group's area of operations). They explained to Sibert the importance of the mission, emphasizing, Lansdale wrote, that "it would be necessary that we act with the utmost secrecy and greatest dispatch" in order to beat the Soviets to the material. Sibert, however, was "very perturbed" at the proposal, according to Lansdale, and "foresaw all kinds of difficulties with the Russians and political repercussions at home." He told the group that he would have to clear the mission with the commanding general of the Twelfth Army before he could agree to anything. Fortunately for Alsos, that commanding general was Omar Bradley. When he heard of the mission's objectives and Sibert's hesitation, Bradley reportedly told his G-2, "To hell with the Russians," and immediately authorized the plan. Sibert sent them on their way with the necessary letters of authority to all the U.S. field commanders in whose areas Alsos would be operating.[71]

On April 17, the task force proceeded to Calbe, the town that housed the command post of the Eighty-Third Infantry Division. There they met with a Colonel Boyle (either the division chief of staff or the G-2),[72] who directed them to a Captain de Masse, the chief of the division's G-2 section responsible for the interrogation of civilians. Captain de Masse had already been to the Wirtschaftliche Forschungs Gelleschaft plant, located in a small town called Leopoldshall, just under two miles from Stassfurt, and knew the director ("Schultz") and the manager ("Schumann"). The Alsos team and de Masse picked up Schultz and Schumann on their way to the plant, and brought with them a copy of the plant's inventory record collected from Schumann's home. It was fortunate that he had a copy, because when they arrived at the plant they discovered it had been badly damaged by both Allied bombing and looting from French and Italian workmen. "The records," Lansdale noted, "were hopelessly strewn about the place." With the manager's record, however, the mission was able to discover approximately 1,100 tons of uranium ore.[73]

The material was in barrels stored in aboveground sheds and, according to Lansdale, "had obviously been there a long time, many of the barrels being broken open." The following day, Lansdale left the rest of the task force at the plant "to take inventory and guard the place" and proceeded to the headquarters of the Ninth Army to arrange to have two truck companies assigned to Alsos to transport the material to Hildesheim, the nearest railhead within the American zone of occupation. On April 19, Lansdale returned to the plant and began to coordinate the transfer of the materials. Many of the barrels, however, were broken and others were in such a weakened condition that they could not be transported. Therefore, Lansdale, along with Bullock and Hambro, located a paper bag factory in the area and confiscated ten thousand "large heavy bags" in which the uranium would be transported. By that evening, the material was repacked and heading for Hildesheim. Lansdale had sent Calvert ahead to receive the material, and by the end of the month the uranium was on its way to Great Britain and then the United States.[74]

In the meantime, German resistance in the South had begun to deteriorate so quickly that the French had been moving much more rapidly than expected. On April 21, the Americans discovered that the French had pushed beyond the line at which they had been ordered to halt, and were moving rapidly toward the target cities (apparently the French were intent on getting to the town of Sigmaringen, where the Vichy French government was located). Colonel Pash, who had returned to Sixth Army headquarters following the completion of the Stassfurt operation, acted immediately. Gen. Jacob Devers, commanding general of the Sixth Army Group, gave Pash operational control of the 1269th Engineer Combat Battalion,[75] and he quickly set off for the first target city, Haigerloch.

The Alsos Mission, with the assistance of the combat engineers, captured Haigerloch on April 23, in advance of the French. According to Pash, as the engineers "were busy consolidating the first Alsos-directed seizure of an enemy town," he sent investigative teams throughout Haigerloch to locate the German research facilities. They discovered a secret German laboratory in a cave "in the side of an 80-foot cliff towering above the lower level of the town," an "ingenious set-up" that gave it almost complete protection from both aerial reconnaissance and bombing.[76] In the cave Alsos discovered a German experimental reactor—an atomic pile—that had been brought there from the Kaiser Wilhelm Institute of Physics in Berlin in February. The pile was equipped with a graphite moderator but did not have any uranium in it. The next day a British scientific intelligence team, escorted by Lansdale and Furman, arrived to help the Alsos scientists evaluate and analyze the reactor. The British included Sir Charles Hambro, Michael Perrin, and David Gattiker from the Stassfurt operation, but also Cdr. Eric Welsh, of British scientific intelligence, and Wing Commanders Cecil and Norman, both of British Secret Intelligence.[77] The scientists measured the pile and quickly determined that, according to Lansdale, it was "simply not big enough" to have been self-sustaining.[78]

Hambro agreed to take responsibility for the dismantling of the pile, so Lansdale and Pash moved on to Hechingen. Pash had already sent the task force ahead of him, and they had captured the town nearly unopposed. The primary target in Hechingen was an old wool mill that now housed the Kaiser Wilhelm Institute for Physics. Within fifteen minutes of their arrival, the Alsos team secured the mill/institute and established a command post. Quickly they began to capture some of the key scientists on their target list, including Carl von Weizsäcker, Erich Bagge, and Karl Wirtz. They thought they would also find Werner Heisenberg at Hechingen, but they learned from their captives that he had left two weeks earlier via bicycle to join his family in the small town of Urfeld in the Bavarian Alps.

The following morning, Pash led a reconnaissance team into Tailfingen, where they captured a large chemistry laboratory and took into custody Otto Hahn and Max von Laue. Hahn agreed to give them all of his secret reports and documents on the entirety of the German atomic bomb program, and they confirmed what Alsos had known since Strasbourg: the German program barely existed. On April 26, Lansdale, Welsh, and Perrin interrogated Laue, Weizsäcker, Wirtz, and Hahn. The Americans and British, Lansdale reported, were "particularly interested in finding the heavy water and the uranium oxide which must have been used in the Haigerloch pile." After a long session of questioning, during which the German scientists denied all knowledge of where the material was located, Karl Wirtz finally agreed to show Alsos the hiding place of the heavy water and uranium. The heavy water was in steel barrels in an old mill about three miles from Haigerloch, while the uranium had been buried in a field on a hill overlooking Haigerloch. Both materials were collected and sent on trucks to Paris for later shipment to Great Britain and the United States.[79] The next day, April 27, the German scientists were sent to Heidelberg for further interrogation. Before they left, Weizsäcker told Alsos that he had hidden his secret papers behind his house. Sam Goudsmit, who by that time had caught up with the mission, fished the papers out of a cesspool on Weizsäcker's property. Enclosed in a metal drum, these papers were a complete set of German atomic bomb documents (and as an added bonus, the papers also included a large secret report on German guided missiles).[80]

With the exception of Heisenberg, Walther Gerlach, and Kurt Diebner, Alsos had captured every significant German atomic scientist. The German atomic pile and all of the remaining fissile material were in American or British hands, and all related equipment and documents had been kept away from the French and Soviets. In addition, the entire operation had been conducted while only twelve hours ahead of the forward advance of the French Army. As Alsos moved to Tailfingen, French Moroccan troops were entering Hechingen. The same French force entered Tailfingen the day after it was captured by the Alsos Mission.[81] By the end of April, as Groves attested, "Alsos was heavily engaged in mopping-up activities." With the majority of scientists captured, and with the fissile material and secret documents secured, "our principal concern at this point," Groves said, "was to keep information and atomic scientists from falling into the hands of the Russians."[82] This would mean one final mission for Alsos.

To the Finish

On April 28, the entire Alsos Mission contingent returned to their Heidelberg base (Alsos Forward South) to plan and prepare for their final operation. It was thought that Gerlach and Diebner were most likely in the vicinity of Munich, while Heisenberg was in Urfeld. Pash decided to split Alsos into two task forces. One, commanded by Major Ham, would proceed to Munich to hunt Gerlach and Diebner. The other, commanded by Pash, would go after Heisenberg. Both groups left Heidelberg the morning of April 30.

Ham's Munich operation included, among others, Capt. Reg Augustine, Carl Baumann (an Alsos scientist), three CIC agents, and three enlisted drivers. On May 1, the Alsos group entered Munich at ten thirty in the morning and proceeded to make contact with U.S. forces. That afternoon, Ham and Baumann went to the home of the first target, Walther Gerlach. Gerlach was not at home, but his wife accompanied the group to the University of Munich, where the physicist was located. Gerlach was found in the basement of the Physics Laboratory, seized, and taken back to his house for interrogation. From this questioning, Alsos discovered the location of its second target, Kurt Diebner. The next day, May 2, Ham's team located Diebner in the town of Schongeising, approximately twenty miles southwest of Munich, and brought him under guard to join Gerlach back in Munich. On May 3, the two German scientists, along with their personal documents, were evacuated from Munich back to Alsos Forward South at Heidelberg. In all, the Munich operation was "a rapid and successful one," Ham concluded. "Personnel targets of interest to the Mission were secured and evacuated according to plan."[83]

The Alsos contingent commanded by Pash had a much more difficult time capturing Heisenberg. The town of Urfeld lay within the area of the "Bavarian Redoubt," where the fanatical, true-believer Nazis were supposed to make their last stand. It was not yet in Allied hands. Pash, however, had been hunting Heisenberg for a year and a half by this time. He was not about to let the fact that Heisenberg was behind German lines, twenty miles ahead of the advance elements of the Seventh U.S. Army, stop him from capturing his ultimate prize. On May 2, Pash and his team approached Urfeld and discovered that the bridge to the town had been destroyed and no vehicles could get through to the city. Deciding to dismount his eleven-man force,[84] he rounded up another ten men from a reconnaissance patrol to move by foot across the mountains into Urfeld. They took the town, without resistance, around 4:45 p.m. An hour later, according to the mission report, "a small force of Germans" attempted to enter the town, but they were repelled by Pash and his unit. The report indicates that the Germans lost "two men killed, three wounded, and fifteen prisoners."[85]

That evening, a German general came to see Pash and attempted to surrender his entire division to Alsos. Pash told him that the general would have to wait until the morning, since Pash did not want to wake up his commander, who was right behind him with a larger force. The German general bought the story, but just as he left the command post, a second German commander entered and also attempted to surrender his forces to Pash. According to Pash, this commander (whose rank is not clear from the documents) indicated to Pash "that there was a force of approximately 700 men in the surrounding mountains" ready to give themselves up to the Americans. Pash was a bold and courageous officer, but even he knew that his force of twenty-one stood no chance of survival once the Germans discovered his true strength, and to remain in Urfeld, he said, "would have jeopardized the execution of the mission." Thus, "after bluffing the Germans in an indication of force, the Alsos unit withdrew on foot to its starting point" and returned to their vehicles.[86]

That night, the bridge to Urfeld was repaired by American combat engineers, and the next day Pash returned to Urfeld supported by an infantry battalion of the 142nd Infantry Regiment. As the town was being secured by the army, the Alsos Mission found Heisenberg in his office, bags packed, waiting to be captured. He was immediately sent back to Heidelberg to join the other German scientists.

Pash reported the capture in his dryly worded mission report: "The personality target was picked up and evacuated."[87] Leslie Groves was more expositive in his analysis of the Urfeld operation. "Pash's last effort," he wrote, "typified the boldness with which he carried out every one of his operations, and clearly demonstrated his ability to stick to his objective, which, in this case, had been to catch Heisenberg. Heisenberg was one of the world's leading physicists and, at the time of the German break-up, he was worth more to us than ten divisions of Germans. Had he fallen into Russian hands, he would have proven invaluable to them."[88]

The war in Europe ended five days after Heisenberg's capture. The German surrender, Pash later recalled, "had thrown wide the gates to [Alsos] scientists, whose interests remained intense in research centers, document centers, laboratories and other places where research could have been carried out." Alsos teams were sent throughout Europe to secure loose ends and to ensure that nothing was left for the Soviets. According to Pash, they were active "in all parts of Germany, Austria, Czechoslovakia, Italy, and, as guests, in Holland, Belgium, and France."[89] However, against the wishes of many within the intelligence field who saw the true merit in an organization such as Alsos (see chapter 5), the Alsos Mission was broken up shortly after the end of the Second World War. "The 144 men and women who were with the mission on V-E Day (28 officers, 43 enlisted men, 19 scientists, 5 civilian employees and 19 CIC agents)," Groves wrote, "were gradually reduced by attrition until, on October 15, 1945, the 'MED Scientific Intelligence (Alsos) Mission' was officially disbanded."[90]

The only outstanding question by the summer of 1945, therefore, was what to do with the German scientists. Groves did not want them to come to the United States, where they "would inevitably learn a great deal about our work and would not for some time make any contribution in return." More importantly, he did not want them to come under Soviet control. With their background, they would be of significant utility to the Soviets. Ten of them (Erich Bagge, Kurt Diebner, Walther Gerlach, Otto Hahn, Paul Harteck, Werner Heisenberg, Max von Laue, Carl von Weizsäcker, Karl Wirtz, and Horst Korsching) were therefore sent to England and secretly detained at an estate in Farm Hall, fifteen miles from Cambridge, while Groves, the American authorities, and the British decided what to ultimately do with them.

From July through December 1945, the scientists' conversations were clandestinely taped, and some of those conversations, most notably between Heisenberg and his colleagues, confirmed Groves's worst fears. On several occasions, Heisenberg was heard telling his colleagues that if the British or the Americans did not intend to allow him to do what he called "proper physics" in Germany, or if the living conditions in Germany were subpar, he would consider working with the Soviets.[91] On another occasion, Heisenberg was heard discussing the potential lure of working in the Soviet Union: "But if in a year or six months' time we find that we are only able to eke out a meagre existence under the Anglo-Saxons, whereas the Russians offer us a job for say fifty thousand roubles, what then? Can they expect us to say: 'No, we will refuse these fifty thousand roubles as we are so pleased and grateful to be allowed to remain on the English side.' "[92]

To prevent defections, the Americans and the British decided that the only prudent solution was to return the scientists to western Germany, but to ensure that, according to Groves, the working conditions there for them "would be such that they could not be tempted by Russian offers."[93] On December 22, 1945, the scientists were notified that they were going to be sent back to Germany. The Americans and the British had spent the better part of the summer and fall constructing and improving laboratory facilities in their zones of occupation so that Werner Heisenberg and the rest of the German atomic scientists would feel content in their working environment. To their credit, this effort achieved its purpose. "Not a single one of these men," Groves later wrote, "left for the East despite the quite attractive offers they must have received from the Soviet Union."[94]

5

Regression

The Postwar Devolution of U.S. Nuclear Intelligence

Discussions about the nature of postwar U.S. intelligence began almost a year before the Second World War ended. In October 1944, Office of Strategic Services director William Donovan met with President Roosevelt to recommend a permanent, centralized intelligence agency placed under the direct supervision of the president. Donovan understood that the OSS was created as a wartime agency designed to support the military directly, and was thus placed under the control of the Joint Chiefs of Staff. His new peacetime agency, he argued, should focus on national and not just military intelligence. The executive branch, with the assistance of both the War and Navy Departments and the secretary of state, should coordinate the new organization. Roosevelt, who had come to trust Donovan's experience and insight, agreed in principle with Donovan's plan, but Roosevelt's death on April 12, 1945, put the OSS director's proposal in jeopardy.[1]

Roosevelt had protected Donovan from much of the bureaucratic infighting that characterized the relations between U.S. intelligence agencies during the Second World War. Army and Navy Intelligence, the Department of State, and the Federal Bureau of Investigation had formed intelligence organizations long before the start of the war,[2] and only Roosevelt's favor had kept the upstart OSS on relatively equal footing. Roosevelt's death, and Harry Truman's rise to the presidency, meant that Donovan would have to fight the parochial interests of each of these agencies without his powerful patron. Truman, as he later explained in his memoirs, was open to the idea of a "sound, well-organized intelligence system," and he agreed that "plans needed to be made." But he argued that "it was imperative that [the United States] refrain from rushing into something that would produce harmful and unnecessary rivalries among the various intelligence agencies."[3]

The Dismantling of U.S. Wartime Intelligence

On September 20, 1945, President Truman signed Executive Order 9621, officially terminating the Office of Strategic Services and spreading its functions throughout the government. The executive order transferred the research and analysis functions of the OSS to the Department of State and the operational functions to the Department of War. The secretary of state was given the power to choose what parts of the research and analysis branch of the OSS (at the time known as the Interim Research and Intelligence Service) he thought could benefit the State Department and dispose of any other personnel, materials, records, or funds he deemed unnecessary. The portions of the OSS the secretary of state would decide to keep would become the Department of State's Office of Research and Intelligence.[4]

The OSS operational units placed under the secretary of war would be organized into an agency called the Strategic Services Unit (SSU). Brig. Gen. John Magruder, the former deputy director of intelligence for the OSS, was appointed as its director. Magruder would be forced to downsize the manpower of his unit, a natural consequence of the end of combat operations, particularly when so many members of the U.S. military in the Second World War were drafted or had only enlisted for the duration of the conflict. Yet John McCloy, the assistant secretary of war, insisted that in the course of the downsizing, the institutional knowledge of the

OSS "must be preserved so far as potentially of future usefulness to the country."[5]

This was no easy task, as the natural attrition of peacetime would begin to take its toll. As of September 30, 1945, the OSS operations units maintained a force of 10,390 personnel—5,713 overseas and 4,677 in the United States (6,964 army personnel, 734 navy personnel, and 2,692 civilians).[6] Of these, 9,058 were transferred to the SSU on October 1. By October 19, that number had been reduced to 7,640, and nearly 3,000 of that figure were in the process of separation. At the end of October, Magruder estimated that overall SSU personnel strength would be further reduced to 1,913 by December 1.[7]

On January 22, 1946, Truman took the first step in the process that would eventually lead to the creation of the Central Intelligence Agency. He designated the secretaries of state, war, and the navy, along with a personal representative of the president,[8] as the National Intelligence Authority (NIA). He directed that "all Federal foreign intelligence activities be planned, developed and coordinated" by the NIA "so as to assure the most effective accomplishment of the intelligence mission related to national security." Within the limits of available funding, the NIA members would "from time to time assign persons and facilities from [their] respective Departments" to collectively form a Central Intelligence Group (CIG). The CIG would be headed by a director of central intelligence (DCI), appointed by the president, who would be "responsible to the National Intelligence Authority, and shall sit as a non-voting member thereof."[9]

The DCI was tasked with the correlation and evaluation of intelligence relating to national security, and with what Truman called "the appropriate dissemination within the Government of the resulting strategic and national policy intelligence." The CIG and the DCI were given no independent budget, personnel, or collection capabilities of their own; instead the CIG would operate with personnel and facilities borrowed from the participating departments, and the director would be required to obtain approval from the NIA for nearly all decisions regarding intelligence functions. In addition, the existing departmental intelligence agencies would continue to collect, analyze, and disseminate "departmental intelligence." Finally, the DCI would be advised in his role by an Intelligence Advisory Board, which would consist of the heads of the departmental intelligence agencies, or their representatives. There was very little that was independent or centralized with this new intelligence organization.[10]

On January 23, the day after Truman's directive, Rear Adm. Sidney Souers was appointed the first DCI. Souers had been the assistant director of the Office of Naval Intelligence during the final eighteen months of the war. He became the director of an organization that had been given general functions and principles, but these remained only broadly defined. Souers began to formulate a plan for the development of the CIG's capabilities for intelligence on the Soviet Union, but this was easier said than done. The Soviets were considered a "hard target." There was little available human intelligence collection on the Soviet problem, the Soviet Union was a closed-off police state with a robust counterintelligence apparatus, U.S. intelligence had very few Russian speakers, and Soviet spies in the West (such as the infamous British intelligence officer Kim Philby) provided early warning of any attempt to infiltrate assets behind the Iron Curtain. To make matters even more difficult, this was a time before the United States could take full advantage of the technological genius that would come to define Cold War espionage—the fully realized signals intelligence capabilities of the National Security Agency were still years away, and it would be another decade before imagery intelligence from spy aircraft (like the U-2) or satellites (like Corona) became available.

On April 29, Souers sent a memorandum to the NIA explaining that, based on the NIA's informal concurrence, a Planning Committee had been formed "to utilize the facilities of all interested Government agencies for the production of the highest possible quality of intelligence on the U.S.S.R." Consisting of representatives from the CIG, the State Department, the Military Intelligence Service (G-2), the Office of Naval Intelligence, and A-2 (Army Air Force Intelligence), the committee drew up a plan to "coordinate and improve the production of intelligence on the U.S.S.R." Acknowledging the "urgent need" to develop actionable intelligence on the Soviet Union "in the shortest possible time," the committee established a Working Committee tasked with producing a compilation of known strategic intelligence on the Soviet Union. These Strategic Intelligence Digests would then be distributed to the member agencies and used to create Strategic Intelligence Estimates "as required to meet [the agency's] needs and also whenever requested by the Director of Central Intelligence."[11]

A year later, on July 26, 1947, Truman signed the National Security Act of 1947 into law.[12] This legislation created an independent air force and united the military services under a secretary of defense. In addition, it established the National Security Council, a body designed to centralize and coordinate national security policy in the executive branch. A final key provision of the act was the formation of the Central Intelligence Agency, the United States' first centralized, independent intelligence organization designed to be the primary information clearinghouse for the National Security Council and the president.[13]

While in theory the National Security Act of 1947 and the creation of the CIA should have solved many of the issues the intelligence community had faced since the end of the Second World War, it did not immediately fix some of the most pressing problems. The CIA experienced growing pains before it became an effective centralized intelligence organization. Some of these were the same problems that intelligence officials had faced under the CIG—it was hard to spy on the Soviet Union.

U.S. scientific and nuclear intelligence followed a path that in some ways mimicked the fall and rise of the broader intelligence system. But in key ways, this intelligence presented a series of its own extremely unique—and uniquely difficult—problems.

The Decline of U.S. Scientific Intelligence

Scientific intelligence in the postwar United States was severely hampered by a critical shortage of qualified scientists in government service. When the war ended, many of the nation's scientists left government service and returned to their civilian careers. Most of the top scientists went back to their academic posts, while many junior scientists left to complete their advanced degrees or to begin their academic careers. Universities were willing to hire junior government scientists, particularly those who had worked on the atomic bomb, at positions far higher than they could have received before the war.[14] Industry would also play a detrimental role in manpower shortages. Scientific research and development had become lucrative during the war, and thus major U.S. industrial firms were competing for the services of experienced scientists.

Complicating matters further, the lack of suitably trained personnel limited the government's ability to recruit new scientists to government service. During the war, the draft board did not give deferments to science students, even those in graduate school. Vannevar Bush, in a report to Truman in 1945, estimated that the war prevented 150,000 potential scientists from obtaining their bachelor's degrees.[15] The war also kept close to 10,000 scientists from earning their doctoral degrees (a number equivalent to all scientific PhDs granted in the United States between 1898 and 1927).[16] By 1955, Bush argued, the shortfall in scientific doctoral degrees would be close to 17,000. Since 1940 and the passage of the Selective Service Act, he noted, there had been "practically no students over 18, outside of students of medicine and engineering in Army and Navy programs, and a few 4-F's, who have followed an integrated scientific program in the United States."[17] According to Bush, because Selective Service policies did not take into account the "Nation's vital needs" for scientists,[18] the United States entered the postwar period "with a serious deficit in our trained scientific personnel."[19]

The reasons for the exodus of scientists from government service were varied. During the war, most scientists came into government because of a sense of patriotism, or at least a belief that the Axis needed to be defeated at all costs. As science became centralized under the OSRD and the MED, university laboratories were commandeered for national research, and scientists temporarily ignored questions of patents and individual achievement for collective effort. In order to help the Allies win the war, the scientists were willing to endure the regimentation of government science, the loss of personal freedoms, the inability to continue their personal research, and the de-emphasis on basic science. That goal accomplished, they were ready to return to private life. Some wanted the ability to resume publishing their scientific discoveries, something governmental security policy prevented them from doing. Others were swayed by the lure of higher salaries in industrial laboratories, or by the opportunity to move into executive and administrative positions (with even higher salaries). Government science restricted the number of personnel who could reach management, and still those positions offered much lower salaries than did industry.[20]

A major concern for American scientists was the state of basic science after the war. For its entire history, the United States had looked to Europe for leadership in basic science, and this trend continued up until the beginning of the war (see chapter 1 for an analysis of the European, and especially German, contribution to the sciences). With the infrastructure of European science in ruins due to the war, the United States would be forced to develop its own foundation of basic science in universities, industry, and private institutions (such as Carnegie or Rockefeller). American scientists believed that the fundamental scientific knowledge developed in the decades prior to the war had been exhausted, and that only a concerted effort to make up for this loss could put American postwar science on a firm footing. This would require a return to private life.[21]

These incentives pulled scientists from government service back into universities and industrial laboratories. However, there were perhaps more powerful factors that pushed American scientists away from government work. Anti-Communism and anti-intellectualism, later personified as McCarthyism, severely embittered the relationship between science and the government. Soon after the end of the war, prominent American scientists were subjected to illegal surveillance by the FBI, interrogation from the House Un-American Activities Committee (HUAC), accusations by the media that they were Communist sympathizers or spies, and federal indictments for disloyalty. According to the MIT physicist and historian of science David Kaiser, "The early years of the Cold War were not a pleasant time to be an intellectual in the United States."[22]

Theoretical physicists were hit particularly hard by this Cold War hysteria. HUAC publicly accused more than a dozen theoretical physicists of Communist infiltration of weapons projects and educational institutions. In most cases, these scientists had close ties to Robert Oppenheimer, who became the most public face of the abuse of scientists by the U.S. government.[23] Loyalty oaths became necessary for government work, which alienated even more scientists, and the perception of guilty-until-proven-innocent became a normal part of a scientist's life. Even those who wanted a career in government science were faced with significant difficulties. Because of suspicion by the government in the early Cold War era, government scientists had problems receiving security clearances for their secret work. By 1949, the backlog of clearance applications had reached a critical level. That year the *New York Times* reported that "somewhere between twenty thousand and fifty thousand scientists, engineers, and technicians" had not been cleared for government employment by the FBI.[24]

The hostile relationship between science and the government became so acute in the late 1940s that Truman felt he needed to address it directly. In a speech he gave before the American Association for the Advancement of Science on September 13, 1948, the president acknowledged that "it is highly unfortunate that we have not been able to maintain the proper conditions for best scientific work. This failure has grave implications for our national security and welfare." Scientists, he said, were discouraged from working in government because they "want to work in an atmosphere free from suspicion, personal insult, or politically motivated attacks." Truman declared that the situation was of particular concern for him, and cited a telegram he received "from eight distinguished scientists." In the telegram the scientists "expressed their alarm" at the state of the relationship between science and the government, "because of the frequent attacks which have been made on scientists in the ostensible name of security." The security environment, as it then stood, made scientists "shun Government work," and they were reluctant to work where they would be open to "smears that may ruin them professionally for life." The indispensable work of government science "may be made impossible," Truman said, by the atmosphere of rumor, gossip, and vilification. To him, "such an atmosphere is un-American, the most un-American thing we have to contend with today." He continued, "It is the climate of a totalitarian country in which scientists are expected to change their theories to match changes in the police state's propaganda line." The government could not force scientists back into government service, but, Truman argued, if this conduct were to continue, "if we tolerate reckless or unfair attacks, we can certainly drive them out."[25]

Recruiting and retaining qualified scientists for government work was not the only obstacle facing the establishment of an effective scientific intelligence apparatus during the early Cold War period. Equally problematic was the indecision regarding the place of scientific intelligence within the broader intelligence community. The intelligence leadership in the United States could not decide what agency should be in charge of scientific intelligence, and no agency wanted the responsibility of organizing such an embryonic and ill-defined field. As a result, scientific intelligence was relegated to the status of an afterthought in the late 1940s. On January 2, 1947, the National Intelligence Authority, in NIA Directive No. 7, "Coordination of Collection Activities," created the policies and objectives to govern "interdepartmental coordination of [intelligence] collection activities so that measures may be taken promptly to effect sound and efficient utilization of the various departmental overseas collecting and reporting activities." That is to say, the NIA assigned responsibility to different agencies for different fields of intelligence collection. Political, cultural, and social intelligence would be the responsibility of the State Department. Military intelligence would reside in the War Department, while naval intelligence naturally was the domain of the Department of the Navy. Scientific intelligence, however, was assigned to "each agency in accordance with its respective needs." Although the NIA directive did mandate that intelligence material, regardless of what agency did the collecting, should be transmitted immediately to a representative of the agency "most concerned" with the information, the directive did not define the agency that should be most concerned with scientific intelligence.[26]

The Central Intelligence Group was not included in the NIA directive, most likely because the CIG was not designed to be primarily an intelligence collection organization. With the creation of the CIA in 1947, one might assume that this new centralized organization, capable of intelligence collection in its own right, would then be part of any broader scientific intelligence collection effort. Instead, however, the National Security Council (NSC) codified the NIA's earlier allocation. In NSC Intelligence Directive No. 2, "Coordination of Collection Activities Abroad," and No. 3, "Coordination of Intelligence Production," both released on January 13, 1948, the NSC reproduced NIA Directive No. 7 almost verbatim—the only difference was that air intelligence was given to the newly created air force.[27]

By the beginning of 1949, the U.S. government took steps to try to improve the coordination and production of scientific intelligence. On December 31, 1948, General Order No. 13 established the Office of Scientific Intelligence (OSI) within the CIA.[28] According to a "Statement of Functions" issued in February 1949, the OSI was designed to be "the primary intelligence evaluation, analysis and production component of CIA with exclusive responsibility for the production and presentation of national scientific intelligence." It was tasked with the preparation of intelligence reports detailing the scientific progress of foreign nations, the review of basic scientific intelligence produced by other agencies, participation in the formulation of the "National Scientific Intelligence Objectives," the guidance of collection efforts, the assisting of interagency committees to aid coordination of effort, and the advising of the director of central intelligence "on programs, plans, policies and procedures for the production of national scientific intelligence." The OSI had its own administrative and analytical staff, and was headed by an assistant director for scientific intelligence.[29]

Less than three weeks later, the NSC released Intelligence Directive No. 10, "Collection of Foreign Scientific and Technical Data." This directive acknowledged the ambiguity of the previous NIA and NSC guidelines for scientific intelligence collection, and attempted to assign certain fields within scientific intelligence to specific agencies. The State Department was given the primary responsibility for the collection of information in the basic sciences "for all government agencies." The Departments of the National Military Establishment (the army, navy, and air force) would collect scientific and technical intelligence for their own requirements, "utilizing whenever practicable the facilities of the Department of State for collection in the basic sciences." The CIA, through its director, was responsible for determining which countries should be targeted for collection (in collaboration with everyone else). Finally, each department was responsible for taking the "appropriate measures to obtain the necessary funds from the Congress or from the agencies served" in order to carry out this task.[30]

Unfortunately, neither the formation of the OSI nor NSC Intelligence Directive No. 10 provided the capabilities to address many of the existing problems. While the OSI helped to coordinate scientific intelligence within the CIA, it did very little to facilitate cooperation between agencies. Nothing done internally within the CIA would make the various intelligence organizations more willing to share information and to put aside their parochial interests. The NSC clarified which agencies would be responsible for what kinds of specific intelligence, but in doing so it institutionalized the decentralization of scientific intelligence. Thus, as the United States entered the year that would prove to be pivotal for Cold War geopolitics, it still lacked a coordinated and effective scientific intelligence apparatus.

Atomic Intelligence

Atomic intelligence is a subset of scientific intelligence, and in many ways its postwar problems parallel those detailed above. However, there were significant issues that were unique to atomic intelligence that require a separate analysis. For one thing, there was a lack of consensus within U.S. leadership as to how strenuously U.S. intelligence should target the Soviet atomic weapons program in the immediate postwar period. Some officials, mainly within the military and the intelligence community, advocated a strong and concerted policy that utilized much of the wartime infrastructure in the field of atomic intelligence built and developed by the MED. Others, mostly scientists but also many diplomats, vehemently promoted the internationalization of atomic energy and atomic weapons, and argued against a world where clandestine atomic intelligence would be required. As a result, the development of an effective atomic intelligence organization was prevented by the inability to create a unified, coherent policy on the nature of atomic use in the first years of the Cold War.

Even before the Second World War officially ended, the Research and Analysis Branch of the OSS was urging William Donovan to push for a strong postwar atomic intelligence program. In a memorandum dated August 18, 1945, Donovan was told that with the invention and use of the atomic bomb, "the nation that pays [the] most intelligence attention to this problem will benefit greatly and . . . any nation which allows itself to be lulled into inattention to these problems will suffer."[31] On September 4, the officer in the OSS who "handled the details of Azusa matters" in the agency concurred with the earlier memorandum and argued that plans should be made "to gauge atom power development by scientists of all countries." The OSS, or whatever agency replaced it, he said, should seek information "regarding [the] scope of their work and their results as there will be strenuous and thorough investigations to develop substitutes for the uranium atom bomb, and for the use of atoms in the development of power."[32]

The navy leadership was getting much of the same advice. On September 22, Capt. William D. Puleston, the former director of naval intelligence,[33] wrote a letter to Adm. Frederick Horne, vice chief of naval operations, arguing for a centralized intelligence agency with powerful atomic intelligence capabilities. In his twenty-two-point proposal to Horne, Puleston wrote that the failure to act on available intelligence had led to the "fiasco" of Pearl Harbor. In the same way, he contended, the inability of current U.S. intelligence to know the atomic capabilities of an enemy meant that there was "little use to maintain either an intelligence service or a Navy, because the enemy can, by a surprise attack in the future, lay waste American industrial cities and probably deal an irreparable blow to our fleet." He called for a centralized agency that would work with the departmental intelligence agencies to "establish secret agents abroad who will endeavor to ascertain the rate of progress of foreign nations in developing any new weapons and their intentions, friendly or hostile, towards the United States." Specifically, Puleston singled out the Soviet Union as the primary threat to the United States. He argued that it was the most likely country to develop atomic weapons in the near future, and thus "the necessity of knowing whether or not Russia is manufacturing atomic bombs is of such importance that immediate measures should be taken to ascertain this fact." For Puleston, these immediate measures included a penetration of the Soviet Union by young American scientists, perhaps of white Russian descent, who would be willing to live and work in the Soviet Union for many years in order to work their way into the confidence of the Soviet scientific community. Possibly some young Polish, Latvian, Finnish, Lithuanian, or German scientists who would "hate the Russians enough" could join the American effort. Whatever the means, something had to be done. All the departments of the U.S. government had habitually neglected foreign intelligence, Puleston argued, but the advent of the atomic bomb would "compel the United States government to revise its attitude or to live in constant jeopardy."[34]

On November 14, 1945, William H. Jackson—an intelligence officer under Gen. Omar Bradley in the Second World War, and the future deputy director of central intelligence—wrote to Secretary of the Navy James Forrestal. Jackson observed that "consideration of most subjects starts today with the conjectural effects of the atomic bomb." With that in mind, he argued for a central intelligence agency to coordinate the collection, evaluation, and collation of national intelligence. For Jackson, intelligence was an essential function of national security, and could only be done effectively through a "comprehensive and integrated intelligence system." If the lessons of Pearl Harbor were not sufficient to convince American policymakers of the urgent necessity for coordination of intelligence within the government, "the use of atomic energy and the threat of yet undeveloped products of scientific research must now supply that proof beyond shadow of doubt." The United States must, Jackson insisted, achieve coordination of intelligence functions to create "a common understanding of the capabilities and intentions of potential enemies," and to prevent a future atomic catastrophe.[35]

The leadership of Army Intelligence had an even more aggressive approach to the potential Soviet threat and the lack of atomic intelligence. On August 30, 1945, Maj. Gen. Clayton Bissell, the army's G-2, proposed a plan to George Marshall, chief of staff, for a permanent, worldwide Alsos Mission. With the concurrence of Leslie Groves and the MED, Bissell's proposal called for a reorganization of Alsos to direct it "toward learning whether scientific, technical, and industrial advances in ostensibly friendly countries throughout the world constitute an imminent military threat to the United States." Operating as a small agency under the administrative direction of the chief of the Military Intelligence Service, the permanent Alsos Mission would work with the OSRD and Groves's office to determine collection targets and to secure information from foreign countries in support of OSRD and MED activities. The plan called for a cadre of scientists "familiar with the techniques of military intelligence collection" to be retained as consultants, and for the training of inexperienced scientists in the methodology of intelligence collection and analysis.[36]

Although Marshall did not approve the plan for a permanent, worldwide Alsos Mission, the army proposal demonstrated the determination within Army Intelligence to strengthen U.S. atomic intelligence. Yet one of the primary reasons the plan was rejected, and, in fact, one of the major reasons for the slow progress in atomic intelligence despite those demanding immediate action, was the strong belief within the U.S. scientific community that a secret nuclear arms race between the United States and the Soviet Union should be avoided at all costs.

The first calls for the internationalization of atomic energy began nearly a year before the end of the Second World War and more than nine months before the first U.S. atomic explosion in Alamogordo, New Mexico. On September 30, 1944, Vannevar Bush, chair of the OSRD, and James Conant, chair of the NDRC and Bush's primary deputy, wrote a memorandum to Secretary of War Henry Stimson. Entitled "Salient Points concerning Future International Handling of Atomic Bombs," it warned that the United States and Britain would not be able to maintain their nuclear monopoly indefinitely. Bush and Conant argued that it would be impossible to keep complete secrecy about the science of the bomb, and so the United States should plan to give "complete disclosure of the history of the development and all but the manufacturing and military details of the bombs as soon as the first bomb has been demonstrated." They contended that it would be "extremely dangerous" for the United States and Great Britain to try to develop the bomb in complete secrecy, since "Russia would undoubtedly proceed in secret along the same lines." Thus, in order to avoid a clandestine atomic arms race, they proposed a "free interchange of all scientific information" on atomic bombs, under "the auspices of an international office deriving its power from whatever association of nations is developed at the close" of the Second World War.[37]

Seven months later, the Interim Committee on the Military Use of the Atomic Bomb met on May 31, 1945, to address this issue. The committee, chaired by Secretary of War Stimson, included Undersecretary of the Navy Ralph A. Bard, Assistant Secretary of State William L. Clayton, Truman adviser and future secretary of state James Byrnes, Bush, Conant, Robert Oppenheimer, Enrico Fermi, Arthur Compton, Ernest Lawrence, George Marshall, and Leslie Groves. At the meeting, Oppenheimer joined Bush and Conant in calling for disclosing all information about atomic weapons to the Soviets. He reasoned that fundamental knowledge of atomic physics was so widespread throughout the world that it might be "wise for the United States to offer to the world free interchange of information with particular emphasis on the development of peace-time uses" before the bomb was used against Japan. If the United States did so, he said, "our moral position would be greatly strengthened." Oppenheimer argued that the Soviets had always been "very friendly to science" and that it might be possible to cooperate with them on atomic control. He felt strongly "that we should not prejudge the Russian attitude in this matter."[38]

Arthur Compton agreed as well. He stressed that the United States should work toward the establishment of a "cooperative understanding" with the Soviets. He favored "freedom of competition and freedom of research activity to as great an extent as possible consistent with security and the international situation." He likewise argued that rigid security over atomic science would actually be detrimental to American science in the long run, since it would result in a "certain sterility of research and a very real competitive disadvantage to the nation." The only way to maintain the United States' current technical advantage over other nations, he said, would be by "drawing on the free interchange of scientific investigation and curiosity."[39]

After the atomic bombs were used against Hiroshima and Nagasaki, American scientists became even more determined to influence U.S. foreign policy toward international control. On November 5, 1945, Vannevar Bush wrote Secretary of State Byrnes and reiterated his concerns. Bush wrote, "The objectives are clear. We wish to proceed down the road of international collaboration and understanding, to avoid a secret arms race, and above all to avoid a future war, in which atomic bombs would devastate our cities as well as those of our enemy." Bush conceded that the Soviets were by nature secretive and suspicious, but still he argued that the United States should approach the Soviet Union with the suggestion that the Soviets join the U.S. and Great Britain to create, within the United Nations, a scientific organization "charged with the full dissemination of fundamental information on science in all fields including that of atomic fission." All of this, Bush continued, would be predicated on the formation of a UN-constituted inspection system with scientific and technical specialists from various countries, which would have the right, without impediment, "to visit any laboratory or plant in any country where atomic fission is being carried out, to the extent necessary to determine the magnitude of the operation, the disposition of the product, etc." While fissionable materials could, of course, be shifted to bomb production once the inspectors had left a country, Bush believed that this would take time, "and would be a fairly obvious procedure if it resulted in shutting down large power plants." Thus, if the inspection program was effective, the internationalization of atomic weapons could remove the threat of surprise atomic attack by one nation on another.[40]

Bush was supported by groups of Manhattan Project scientists who had organized nationwide in order to promote international control. The Atomic Scientists of Chicago, the Association of Oak Ridge Scientists, the Association of Los Alamos Scientists, and the Association of Manhattan Project Scientists joined other scientists' organizations around the country in the wake of Hiroshima and Nagasaki to agitate against a secret atomic arms race. On November 30, 1945, these groups came together to form the Federation of Atomic Scientists (later renamed the Federation of American Scientists), and spent the next two years promoting civilian and international control of atomic energy. Most of these scientists were in their twenties and thirties, but they were soon joined by older and more prominent members of the scientific community.[41] Harold Urey, the Nobel Prize–winning chemist and discoverer of heavy water, argued that "we must expect some sort of world government with adequate powers to prohibit atomic bombs. It must have the power to police the world to see that such laws are obeyed." The United Nations was not yet capable of accepting this responsibility, he said, so the United States should attempt "to strengthen that organization in such ways as to make it a more effective world government."[42]

Robert Oppenheimer and Albert Einstein, the two most famous scientists in the United States in 1945, were also vocal and public proponents of international control. Oppenheimer, who had famously told President Truman that he had blood on his hands following the atomic bombing of Japan, told the U.S. Senate on December 5, 1945, that the United States should destroy its stockpile of atomic bombs if that action could result in world peace. He testified to the Senate that there was "a good reason for attempting to establish in the international control of atomic armament [patterns] of confidence, collaboration and good faith which in a wider application must form the basis of peace. There may not be a comparable opportunity again."[43]

Einstein formed his own organization early in 1946. The Emergency Committee of Atomic Scientists consisted of eight scientists who were heavily involved in the creation of the U.S. atomic bomb: Einstein, Leo Szilard, Harold Urey, Hans Bethe, Victor Weisskopf, Linus Pauling, Philip Morse, and T. R. Hogness. Their goal was to warn the public of the dangers of atomic weapons, and in June 1946, Einstein gave an interview to the *New York Times Magazine* in which he argued that "a new type of thinking is essential if mankind is to survive and move to higher levels." The atomic bomb had altered the nature of the world as people knew it, and in the light of this new knowledge, Einstein said, "a world authority and an eventual world state are not just *desirable* in the name of brotherhood, they are *necessary* for survival." In previous periods the strength of a nation's army could protect it from destruction, but in the atomic age, countries must abandon competition and embrace cooperation or the world faced "certain disaster." Therefore, he argued, "every nation's foreign policy must be judged at every point by one consideration: does it lead us to a world of law and order or does it lead us back toward anarchy and death?"[44]

The atomic scientist movement had two major policy priorities. The first was to remove atomic power and atomic weapons from military control. Due in large part to the scientists' passionate lobbying, the U.S. Congress passed the Atomic Energy Act of 1946, more popularly known as the McMahon Act, after its sponsor, Senator Brien McMahon. The McMahon Act went into effect on January 1, 1947, and created the U.S. Atomic Energy Commission (AEC), a civilian agency that was given responsibility for nuclear power development and nuclear weapons development and control. The AEC would be given scientific and technical advice by a General Advisory Committee (GAC), which was made up of seven prominent atomic scientists and two industrialists. The membership of the GAC included Robert Oppenheimer, Enrico Fermi, Glenn Seaborg, James Conant, I. I. Rabi (a nuclear physicist at Columbia), Lee DuBridge (president of the California Institute of Technology), Cyril Smith (director of the Institute for the Study of Metals), Hood Worthington (of DuPont), and Hartley Rowe (of the United Fruit Company). Collectively, the members of the GAC had been involved in every phase of the U.S. atomic bomb project and understood every aspect of nuclear science.[45]

The scientists' second policy consideration was less successful. The Baruch Plan, named after Bernard Baruch, the senior U.S. representative to the United Nations atomic energy negotiations, was the United States' proposal for the internationalization of atomic energy. The plan called for a limited world government that would regulate atomic energy worldwide. Any attempt by a country to build atomic weapons would be punished—and presumably prevented—by the UN. The UN Atomic Development Authority would pursue nuclear power for use by all nations, and would provide for an open scientific world in which scientific research would be free and unrestricted and in which all scientists could work for the betterment of mankind.[46]

With the Baruch Plan, the United States was proposing to give up its atomic weapons arsenal in order to prevent others from developing stockpiles of their own. The Soviets, however, rejected the plan. They could not accept a plan that allowed for the preservation of the United States' geographic and technological advantage while at the same time opening up the Soviet Union to the U.S.-controlled United Nations. Before they would agree to even consider international controls, the United States would have to unilaterally dismantle its atomic bomb program as part of a broader plan to outlaw atomic weapons. Only then would the Soviets be willing to negotiate a control system. Even then, by "control" they meant a system of periodic inspections of national facilities that would limit the authority of the UN so much that it would be essentially powerless to discover Soviet circumvention of the treaty.[47]

No one was happier that the Baruch Plan failed than Leslie Groves. He had been arguing against the idea of international control of atomic weapons since the idea was first conceived. Groves believed that it was naive to think that the Soviet Union would not covertly break any agreement to forgo atomic weapons development. He contended that to follow through on the plan to internationalize atomic weapons "would be to invite disaster to the United States unless [his] lack of belief in the good faith of other nations proves to be without justification."[48] Groves was therefore frustrated by the delay in the creation of an effective national atomic intelligence organization, and he refused to wait until the policy-making community realized the folly of their indecision.

Although the Alsos Mission had been disbanded, and although most of his more experienced officers from the Second World War had been reservists and had by then returned to civilian life, Groves was able to maintain a small office of atomic intelligence specialists in the MED. The Foreign Intelligence Section of the Washington Liaison Office of the Manhattan District was staffed by a handful of intelligence officers, consisting of career Corps of Engineers personnel, several officers and civilians trained in science, and several Counter Intelligence Corps agents trained in investigative procedures. While they depended entirely on information from intelligence collection agencies such as the Strategic Services Unit, the State Department, and British Intelligence, in 1945 and 1946 they were able to pull together bits and pieces of information to create a basic picture of the embryonic Soviet atomic weapons program.

From the State Department the Foreign Intelligence Section learned in November 1945 that the Soviets were studying equipment for atomic bomb manufacturing,[49] and that they had ordered the Czechoslovakian government to provide them with uranium ore from the Joachimsthal mine (by then under Soviet control).[50] From the British the Americans learned that the Soviet physicist Peter Kapitza had sent a secret letter to the Danish physicist Niels Bohr, inviting him to work on atomic fission in the Soviet Union. According to the British diplomat Roger Makins, Kapitza had apparently been instructed by the Soviet leadership to deliver his correspondence "under conditions of absolute secrecy so as to ensure that no other government would have been aware that the meeting had taken place."[51]

The MED intelligence team also learned from the British that several German scientists had gone to the Soviet Union to work in its atomic bomb program. While none of these men were among the top echelon of German atomic scientists (those had been captured by the Alsos Mission), they were competent physicists and chemists who would certainly be a valuable asset to a fledgling Soviet atomic weapons program. Nikolaus Riehl of the Auergesellschaft Plant at Oranienburg brought his entire team to the Soviet Union to help the Soviets produce uranium metal. Gustav Hertz, a Nobel Prize winner in atomic physics and the discoverer of the gaseous diffusion method for separating uranium isotopes, had flown to Moscow in the summer of 1945. Adolf Thiessen, the former director of the Kaiser Wilhelm Institute for Physical Chemistry, had arrived in the Soviet Union that fall, along with eighteen of his subordinates. In addition to Riehl, Hertz, and Thiessen, MED intelligence learned that the Soviets had recruited perhaps a hundred scientists and technicians from Austria and Germany to work on different elements of their bomb program.[52]

By February 1946, the SSU had discovered through agents in the Soviet zone of Germany the locations and activities of many of the Germany scientists. Those working on cyclotron operations were sent to the Crimea in the summer of 1945, and then in October were moved to a more permanent location on the eastern shore of the Black Sea. Thiessen and Hertz were also reported to be located in the Black Sea region, although as late as November 1945, they were still waiting for their housing and laboratories to be built by the Soviets. Nikolaus Riehl's Auergesellschaft group had yet to be located.[53]

The fact that the Foreign Intelligence Section was able to obtain any information about the Soviet atomic program is extraordinary. The lack of an in-house collection capability meant they were forced to rely either on the SSU, a temporary organization with budget and manpower problems; the State Department, widely considered at best a marginal intelligence collector (it had superb analysts but a minor collection capability) and at worst a vocal opponent of the military concerned primarily with its own parochial interests; or the British, who produced excellent intelligence but only provided it to the Americans sparingly, and who could cut off the flow of information at any time.

The Bureaucratic Black Hole of Nuclear Intelligence

At any rate, the formation of the National Intelligence Authority and the Central Intelligence Group in early 1946 should have mitigated many of the MED Foreign Intelligence Section's problems. The CIG was a natural fit for Groves's atomic intelligence organization: a centralized agency under the direction and authority of the military. Both he and Hoyt Vandenberg, the director of central intelligence, advocated its transfer, and both assumed that its integration into the CIG was a foregone conclusion. Yet the passage of the McMahon Act on August 1, 1946, put that plan in jeopardy. The legislation mandated that the Atomic Energy Commission take over all aspects of the Manhattan Engineer District, and David Lilienthal, the chair of the AEC, insisted that this should include the MED's atomic intelligence operation. Groves and Vandenberg challenged this assertion, maintaining that the intelligence functions of the MED were separate from the intended scope of the McMahon Act.[54]

The final arbiter of this dispute between the CIG and the AEC would be the National Intelligence Authority, as only the NIA could approve the transfer of the Foreign Intelligence Section. The NIA was scheduled to have its sixth meeting on August 21, and the fate of atomic intelligence was placed on the agenda. In the weeks before the NIA meeting, the representatives on both sides of the issue cultivated and refined their arguments. Vandenberg and the CIG created a draft NIA directive for the NIA to consider, which specified the DCI as the primary coordinator of the collection of all intelligence information related to foreign atomic energy developments and appropriate dissemination within the government. Additionally, the proposed NIA directive called for the official transfer of the MED intelligence section to the CIG, along with all of its working files and personnel.[55]

On the day of the meeting, NIA secretary James Lay Jr. gave Vandenberg a series of talking points to use while arguing his case.[56] He was told to emphasize the fact that the proposed directive had the concurrence of the permanent members of the Intelligence Advisory Board and Leslie Groves. He was also given three arguments he could use to demonstrate that the CIG plan would not conflict with the McMahon Act and the establishment of the AEC. First, the AEC would deal primarily with domestic atomic energy and weapons developments, while the CIG would deal only with intelligence concerning foreign developments. Second, the CIG would supplement, rather than conflict with, the AEC, and any intelligence the CIG collected would be disseminated, if appropriate, to the commission. Finally, the Foreign Intelligence Section had been considered by Groves as part of his personal staff, rather than an integral part of the MED. The intelligence function, therefore, according to Lay, should not be "involved in the transfer to the Commission of the domestic responsibilities of the Manhattan Engineer District."[57]

The NIA convened at eleven o'clock in the morning on August 21 to make its final determination. Arguing on the side of Groves and the CIG were Secretary of War Robert Patterson, Secretary of the Navy Forrestal, and, of course, Vandenberg. Patterson contended that the Foreign Intelligence Section had nothing to do with the Manhattan Engineer District and therefore had nothing to do with the AEC. He believed that the NIA should have acted long before that day to bring atomic intelligence under its purview, since the unit dealt with what he considered to be military intelligence and fell under the terms of the president's directive to the NIA. He therefore felt the proposed action should be taken immediately. Forrestal concurred with most of what Patterson argued, and stressed that there was no intent to deny intelligence information to the AEC. He added that when the NIA was conceived, it was the intent of the president to draw together all intelligence activities, and not to isolate or separate one unit. He concluded his remarks by stating that "the N.I.A. would be doing a dangerous thing to mark time on this matter."[58]

The lone voice in opposition to the plan was Acting Secretary of State Dean Acheson, who also chaired the NIA meeting. Acheson informed the other members of the NIA that he had spoken to President Truman about the matter and that Truman had expressed to him a desire to wait to make his decision until the full membership of the AEC had been established (it had a chair, but the full commission had yet to be appointed). Acheson was also concerned about the ability of the AEC to discover and acquire foreign sources of uranium ore. If this was something that was intrinsic to atomic intelligence, then, he said, it would be "of vital interest to the Atomic Energy Commission." Acheson was not necessarily against the eventual transfer of the MED's atomic intelligence contingent to the CIG, but because this was such a "complex subject," he was worried about "acting too hurriedly." According to Acheson, the members of the Atomic Energy Commission, once it was fully formed, "should have an opportunity to express their views."[59]

The NIA would finally decide to recommend to the president that he approve the directive to move the Foreign Intelligence Section to the CIG, "with an understanding that any action taken by the N.I.A. will be without prejudice to future change that may be desired by the Atomic Energy Commission."[60] That way, the proponents of the move would get their immediate action, while the potential objections of the AEC could still be addressed at a later date. Adm. William Leahy, the president's chief of staff and representative to the NIA, telegraphed the results of the meeting to Truman, who was away from Washington.[61] The president, however, told Leahy and the other members of the NIA that he wanted to wait until he returned to Washington before making any decision. Upon his return, Truman opted to delay his decision further, choosing to wait until the entire AEC was appointed and confirmed before he would even consider the intelligence issue.[62]

Frustrated by such dithering, Leslie Groves appealed directly to the AEC. On November 21, 1946, Groves wrote the commission a letter outlining why he believed the CIG would be the ideal location for his Foreign Intelligence Section. Groves argued that it was "vital to the security of the United States that foreign intelligence in the field of atomic energy be maintained and strengthened," and that "the CIG must be able to evaluate the capabilities of other nations to use atomic energy in the military field." The best way, he said, to build this capacity within the CIG was "unquestionably" the MED's Foreign Intelligence Section. To continue the functions of MED atomic intelligence in any other way except under the control and direction of the CIG "would be very difficult." The Foreign Intelligence Section had been dependent on the collection capabilities of military intelligence, the SSU, and British Intelligence. Now that the CIG controlled the SSU and had developed close ties with British Intelligence, Groves contended, "it would be a mistake to use the present limited

Manhattan resources based upon informal liaison with the State, War and Navy Departments or any organization set up with the A.E.C." The experience of his section, combined with the mission and operation of the CIG, "logically place them together," but cooperation between the CIG and the AEC would be "absolutely necessary" for national security. The Foreign Intelligence Section, under the control of the CIG, Groves said, "would be the best instrument to provide this coordinated effort."[63]

Whether it was Groves's plea that finally convinced the chair of the AEC may never be known, but by the end of 1946, Patterson, Vandenberg, Lilienthal, and Groves had reached a compromise. The agreement, as it was explained by Patterson to the ninth meeting of the NIA on February 12, 1947, would place the MED atomic intelligence division in the CIG, but would allow three representatives of the AEC access to the MED files set to be transferred to the CIG. The AEC personnel would "search these files for information pertaining to uranium deposits and such information [would be] retained by the Commission."[64] The compromise would be codified in National Intelligence Directive No. 9, "Coordination of Intelligence Activities Related to Foreign Atomic Energy Developments and Potentialities." Enacted on April 18, 1947, the directive was designed to establish once and for all the authority of the director of central intelligence to coordinate "all intelligence information related to foreign atomic energy developments and potentialities affecting the national security, and to accomplish the correlation, evaluation, and appropriate dissemination within the Government of the resulting intelligence."[65]

NIA Intelligence Directive No. 9 was merely a formality, as the CIG had already prepared for the arrival of Groves's intelligence section. On March 29, 1947, the CIG established the Nuclear Energy Group, Scientific Branch, within the Office of Reports and Estimates (ORE). The mission of the Nuclear Energy Group was, among others, to "conduct and coordinate the necessary research and evaluation of intelligence information and intelligence pertaining to the development of nuclear energy by foreign nations," in order to "prepare estimates of the nuclear energy capabilities and intentions of foreign nations for coordination with and incorporation in intelligence of national interest."[66]

The ORE had been created in 1946 to provide policymakers with short- and long-term estimates of a foreign power's intentions and capabilities. Months before the creation of the Nuclear Energy Group, the

ORE released the first U.S. intelligence estimate of when the Soviet Union would manufacture its first atomic bomb. ORE 3/1, released on October 31, 1946, acknowledged that the office's "information relating to this subject [was] meager," but still concluded that it was probable that the Soviets would develop an atomic bomb "at some time between 1950 and 1953."[67]

The ORE had made its best guess based on very little evidence and only "past experience and reasonable conjecture."[68] Now it was getting the intelligence veterans of the Foreign Intelligence Section, and in July it would receive a guaranteed personnel allocation and congressionally mandated funding when Truman signed the National Security Act, transforming the CIG into the CIA. In October, the Scientific Branch was given a highly experienced and competent leader to guide and develop the office. Wallace Brode, who had been recommended for the position by Vannevar Bush, had worked for Bush in the OSRD during the Second World War. Brode had a doctorate in physical chemistry, and had been recruited in August 1944 to work as a special consultant for the Alsos Mission in London and Paris, where he learned the intricacies of scientific and atomic intelligence.[69] After the war, he served as director of the Science Department of the Naval Ordnance Test Station at Inyokern, California, where he continued to formulate plans for scientific intelligence operations.[70]

With the leadership of Brode, the addition of the MED atomic intelligence specialists, and the institutional and organizational backing of the CIA, the Nuclear Energy Group of the Office of Reports and Estimates should have been an elite intelligence section. In reality, however, it did not live up to expectations. There were two primary reasons for this. The first was that, despite NIA Intelligence Directive No. 9, the belief that the CIA should have sole responsibility for atomic intelligence was not widespread throughout the U.S. government. By the end of 1947, the Atomic Energy Commission had formed its own Intelligence Division, which was not forwarding information to the CIA, even when the information had been directly requested by the agency. The State Department was also failing to send pertinent information in a timely manner. The result was a situation in which disagreements over the role and scope of the CIA in atomic intelligence, according to Ralph Clark, director of the Programs Division, had "Brode completely stymied. [The situation was] blocking his attempts to recruit and organize his staff."[71]

The second reason was internal inefficiencies and incompetence within the ORE and CIA. According to a report by Stephen Penrose, an intelligence officer who served in the OSS, later the SSU, and finally the CIA, the ORE produced intelligence that commanded "little respect from the users of such reports in State, Army or Navy." Army Intelligence had told Penrose that it had received no useful additions to its own information "since the [Research and Analysis Branch] of the OSS had been broken up." Most damningly, the army considered its collaboration with the CIA "to be largely a waste of time, particularly as regards Russian matters." According to Penrose, the head of the ORE Russian division "seems content to rest upon his short visits to Russia as sufficient qualification of him as a Russian expert." Penrose singled out Wallace Brode as "one of the ablest men" in the ORE, and emphasized that Brode had been very critical of the "inflexible and unimaginative organizational and personnel policies" of the ORE and CIA.[72]

A key problem Brode faced was the lack of bureaucratic support at the highest levels of the CIA. The CIA's director of central intelligence, Roscoe Hillenkoetter, did not have the same respect for the intricacies of scientific intelligence as did his predecessor, CIG director Hoyt Vandenberg. He allowed other ORE branches to gather scientific intelligence, and did not provide Brode with the resources or authority to force the army, navy, AEC, or State Department to share information with the Scientific Branch.[73]

The conflict between agencies and the lack of administrative support within the CIA, according to the chief of the Intelligence Section, put "Atomic Energy Intelligence in a critical situation,"[74] and forced the director of central intelligence to establish the Joint Nuclear Energy Intelligence Committee (JNEIC) in November 1947. The CIA would provide the chair of the committee and all of its permanent logistical and analytical staff. The remainder of the JNEIC would consist of representatives from the Department of State, the army, the navy, the air force, the AEC, and the Department of Defense's Research and Development Board.[75] By the end of 1947, the JNEIC had taken over the responsibility of estimating Soviet atomic bomb development from the ORE.

The first such estimate was released on December 15, 1947, and would essentially repeat the findings of ORE 3/1, released more than a year earlier. From that point, the JNEIC released estimates of the status of Soviet atomic energy semiannually. Its second report, on July 6, 1948, stated that "no information [had] been received that necessitates changes in the argument of [the December] report." Because of the ineffectiveness of U.S. atomic intelligence, it had remained necessary for the JNEIC to rely on the knowledge of the American, British, and Canadian experiences in atomic energy in order to project estimates onto the Soviet Union. Although U.S. intelligence had received some new information on the Soviet program since the December report that added "somewhat to our knowledge of the scope and details of the USSR's project," it continued to be "impossible to determine its exact status or to determine the date scheduled by the Soviets for the completion of their first atomic bomb." On the basis of the evidence, the JNEIC estimated that the earliest date by which it was "remotely possible" that the Soviet Union might complete its first atomic bomb was mid-1950, although the "most probable date" was mid-1953.[76]

The January 1, 1949, report repeated the estimates of the July 1948 report almost verbatim. That same day, the Office of Scientific Intelligence was formed within the CIA to try to centralize scientific and atomic information (at least within the CIA). The Nuclear Energy Group, which had been temporarily removed from the Scientific Branch and placed in the CIA's Office of Special Operations in March 1948, was reunited with the Office of Reports and Estimates in the newly created OSI. Wallace Brode had resigned as head of the Scientific Branch in October 1948 (for a myriad of reasons, many of which have been detailed in this chapter), and he was replaced by Willard Machle, a medical doctor and former professor of medicine, who became the first director of the OSI. During the first nine months of 1949, Machle did all he could to consolidate the collection, analysis, and dissemination of national scientific and atomic intelligence within the OSI and CIA, but he was equally as unsuccessful as Brode had been before him. The other U.S. intelligence agencies refused to concede their power to the OSI, and Machle struggled to produce effective and up-to-date intelligence on the Soviet atomic bomb program.[77]

The result was the July 1, 1949, estimate of the status of the Soviet atomic energy project. According to the report, the "information now available substantiates" the dates already estimated in the reports of January 1949, July 1948, and December 1947, and the ORE estimate in October 1946: the earliest possible date was mid-1950, while the most probable date was mid-1953. This time, however, the estimate included "new information" that indicated that the Soviets were pursuing one particular method, not specified in the estimate, that would suggest the first Soviet atomic bomb could not be completed before mid-1951.[78] The July 1 report, which set the most probable date for a Soviet bomb four years in the future, was released less than two months before the detonation of the Soviet atomic bomb on August 29, 1949.

By far the clearest demonstration of the dysfunction within the CIA and the ineffectiveness within the broader U.S. intelligence community was the ORE report of September 20, 1949. The report, Intelligence Memorandum No. 225, "Estimate of Status of Atomic Warfare in the USSR," predicted (if that is the right word) a first Soviet bomb in mid-1953 (mid-1950 at the earliest), *twenty-three days after* the detonation of Joe-1. Not only that, but the report's release postdated the United States' discovery of the Soviet bomb by seventeen days, and was released six days after the date (September 14) on which the vast majority (95 percent) of American experts analyzing the data were convinced the Soviets had, indeed, set off an atom bomb.[79]

On the morning of Friday, September 23, President Truman announced to the nation the news that the Soviets had become an atomic power. The United States had discovered the Soviet atomic detonation through a dedicated nuclear detection program called AFOAT-1 (for Air Force Office of Atomic Energy), which had been developed during the late 1940s. AFOAT-1 used specially equipped WB-29s to collect airborne dust from areas around the Soviet Union and test it for radiation and the other chemical and physical by-products of an atomic explosion. Atomic bomb detection is technological intelligence, not scientific intelligence: since the system is detecting an already developed weapon, and not research in the laboratory or in an experimental stage, the discovery of a Soviet atomic test can no longer be considered scientific intelligence, and is therefore outside the purview of this book. That being said, AFOAT-1 and the U.S. atomic detection system demonstrate both the impact of science on the capabilities of intelligence collection and the desperation of the U.S. military leadership to do something to mitigate the U.S. intelligence community's inability to collect information on the Soviet atomic program. In the end, the AFOAT-1 program was a minor intelligence success amidst larger intelligence failures.

Almost immediately after the detection and public announcement of the Soviet atomic bomb, Americans began to assign blame for the intelligence failure. On September 29, Willard Machle, formally the assistant director for scientific intelligence, wrote a memorandum to DCI Hillenkoetter providing a postmortem for what the memo's subject line called the "inability of OSI to accomplish its mission." Machle acknowledged that little had been accomplished toward correcting the inadequacies expressed by two committees tasked with evaluating atomic intelligence, the Dulles and Eberstadt Committees. These inadequacies were particularly highlighted, Machle said, "by the almost total failure of conventional intelligence in estimating Soviet development of an atomic bomb." Machle lamented that "the USSR completed an atomic bomb in half the estimated time required," and admitted there was "a vast area of ignorance of basic scientific research in [the] USSR and Satellite countries." According to Machle, the inadequacies in national scientific intelligence existed because of conditions both inside and outside the CIA.[80]

The conditions outside the CIA that prevented the OSI from correctly assessing Soviet atomic development included the refusal of the departmental intelligence agencies to recognize the CIA as the central coordinating organization in the national intelligence structure. Additionally, Machle noted that the CIA's lack of authority to effect coordination of intelligence activities was due to the domination of the CIA by the departmental intelligence agencies through the mechanism of the Intelligence Advisory Committee. The only remedy, wrote Machle, was to force the departmental intelligence agencies "to recognize the intent of the National Security Act and the authority granted CIA thereunder."[81]

The underlying condition within the CIA that prevented the OSI from accomplishing its mission, Machle argued, was the failure of the collection branches to recognize that they "exist only to provide services" for the analysts and production offices. The Office of Special Operations, the CIA's collection branch, had failed to "discharge its responsibility for covert collection of scientific and technical intelligence." The OSO's deficiencies, according to Machle, included a lack of effective planning of scientific and technical intelligence operations, and a lack of "any mechanism for relating such planning to the needs for national scientific intelligence." While the OSO had its own integrated scientific staff, they were used only in an advisory manner, and this made it impossible for them to effect planning for scientific and technical intelligence operations. Finally, what Machle called "a fallacious concept of operational security" dangerously limited the dissemination of useful intelligence to the OSI and prohibited "technical guidance of operations by informed and competent analysts."[82]

The Central Intelligence Agency had been intended to be the premier source of national intelligence collection, analysis, and dissemination, yet its internal departments could not work together, and the CIA could not work effectively with other U.S. intelligence agencies. The U.S. government would spend the better part of the next four decades trying to fix the deficiencies of its intelligence organization while living with the reality of Soviet atomic power.

6

Whistling in the Dark

The U.S. (Mis)Perception of the Soviet Nuclear Program

The detonation of the Soviet atomic bomb on August 29, 1949, should have ended all questions about the capabilities of science and industry in the Soviet Union. Yet contentious debate about how the Soviets developed an atomic bomb long before the United States expected them to continued for almost a decade. Even after the detonation, the general picture of the Soviet Union "as a basically backward country," in Herbert York's words, did not change.[1]

The unwillingness to accept Soviet capabilities began immediately after the United States' detection of the bomb. Despite the opinion of the majority of scientific and intelligence experts who analyzed the data, Secretary of Defense Louis Johnson at first refused to believe the Soviets had built an atomic bomb. Johnson instead was convinced that a Soviet reactor might have exploded, and so he rejected the intelligence findings. In response, the Atomic Energy Commission assembled a committee under the leadership of Vannevar Bush that included Robert Oppenheimer; Robert Bacher, the former AEC commissioner; William Penny, the director of the British atomic bomb program; and Hoyt Vandenberg. The committee endorsed the original assessment that the Soviets had indeed detonated an atomic bomb, but still Johnson and even President Truman doubted the conclusion. Finally, on September 23, Truman felt he had no choice but to inform the nation that the Soviets had succeeded in building a bomb.[2]

A little less than a month later, on October 17, the Joint Committee on Atomic Energy met to try to understand how the Soviets had accomplished this achievement far earlier than anyone had predicted (at least anyone in a position of power). The committee was made up of four senators and four U.S. representatives, an equal number from each party, and was chaired by Senator Brien McMahon, a Democrat from Connecticut. The committee had called CIA director Roscoe Hillenkoetter to testify to the possible reasons why the United States was taken by surprise (and why the CIA had failed in its mission). Hillenkoetter struggled to justify the estimates the CIA had provided to Congress and the president, arguing that the estimate of the earliest possible date for the Soviet bomb erred by only a few months from what actually occurred. To Hillenkoetter's credit, he warned the committee against assuming the Russian scientists "are dumb or something"—in thinking so, he said, "we are just deluding ourselves." But this line of argument was quickly dismissed by both Hillenkoetter and the committee as they frantically looked for hypotheses to explain the discrepancy between estimate and reality.[3]

Over the course of the lengthy meeting, the transcription of which ran 137 pages, the committee, with the aid of Hillenkoetter, detailed six reasons for the Soviet accomplishment, none of which included innate Soviet scientific ability. Over the next decade, these reasons, whether individually or in some combination, would guide the United States' narrative regarding Soviet science.

The initial culprit for those unwilling to give credit to Soviet science was the so-called Smyth Report. In early 1944, John Lansdale and Leslie Groves had discussed the problems of security following the public revelation of the atomic bomb. They wanted to limit the dissemination of secret information while at the same time declassifying information that was already known, could be discovered by any competent scientists, and had no real bearing on the production of atomic bombs. They also wanted to create a framework for information secrecy, outside of which it would be illegal to operate. That is to say, Lansdale and Groves would release all the information that they deemed acceptable for public consumption, and everything else would be off limits. Groves asked Henry Smyth, chair of the Department of Physics at Princeton University and someone who had not worked on the Manhattan Project, to prepare the report.[4]

The report was completed in July 1945, and was titled "A General Account of the Development of Methods of Using Atomic Energy for Military Purposes." Groves met with the secretary of war, Henry Stimson, on August 2, 1945, to discuss its release. Also attending the meeting was James Conant, Bush's deputy at the OSRD. Conant advocated for its release, stating that without it, "a serious situation may develop," as information was sure to come out about the bomb through various means. Conant stated that the report would give very little to the Russians, and that "anyone could get the information contained in the report with very little money in less than three months."[5]

At first, Stimson was concerned that the Soviets would need the report to help them build their own atomic bomb. He argued that their scientists and their system of government would prevent them from acquiring this information without the report, since "people who lived under oppression cannot be as mentally alert or possess as much initiative as those who live in a land of free press and free speech." In the end, however, Stimson was convinced by Groves's reasoned argument: the United States could either release the Smyth Report, and thus set parameters that all Americans would be legally bound to stay within, or accept the alternative, in which the Soviets would receive thousands of papers published with a great deal more information.[6]

The report was released to the public on August 12, a few days after the atomic bombings of Hiroshima and Nagasaki. It did not contain any details on how to build a functioning weapon. It did not have any illustrations or diagrams that could help the Soviet Union repeat what the Americans had done, nor did it provide any information about the industrial or manufacturing processes that were so integral to atomic bomb development. Upon its release, the Smyth Report received some press coverage, both positive and critical, but it would later become a source of major contention after the detonation of the Soviet bomb. Nearly every article that was published in the days following Truman's announcement of the Soviet bomb contained some mention of the Smyth Report and its "role" in providing the Soviet Union with the information it needed to build an atomic bomb. Despite the fact that it was intended to *prevent* information from reaching the Soviets, and despite the fact that it contained nothing more revelatory than what, according to Groves, "ten graduate students in nuclear science, supervised by one or two extremely able scientists of the type that could be found in any of the major countries of the world," could have compiled in little time,[7] the Smyth Report nevertheless remained the reason, in many Americans' eyes, for the rapid success of Soviet atomic science.

Another argument was that the Soviet Union began its bomb program before 1945, perhaps as early as 1943 or even earlier. This meant that the predictions of the intelligence community were not wrong, at least with regard to how long it would take for the Soviets to bring their program from theoretical work to completion. If they had begun in 1945 after only learning of the bomb's existence from Hiroshima, then the four years it took them to complete their own bomb demonstrated a scientific capability that rivaled that of the United States. If, however, they began their program earlier, then they needed at least two more years, but perhaps as many as six, to develop their weapon. This would mean the perception of the "backward Soviet Union" could remain intact. The *New York Times* was eager to accept this line of argument. On the front page of the *Times* on the day following Truman's announcement of the Soviet detonation, an article appeared that contended that the United States' estimate of the Soviet bomb was based on the "incorrect assumption" that the Soviet Union did not know about the possibilities of the atomic bomb until Hiroshima, and that "it would therefore be unwarranted to assume that Soviet scientists were completely unaware of the military potentialities of fission until 1945, and that they did nothing about it until then." It would be more reasonable, the *Times* argued, "to assume that they had been working on it in secrecy since January, 1939, and that it thus took them ten, rather than four years, to reach the stage of testing their first atom bomb."[8]

Another contention was that the Soviets had learned the secrets of the atomic bomb through espionage—which soon became the conventional wisdom for much of the Cold War, as it is for many today. To be sure, they did learn much about the Manhattan Project and the British Tube Alloys project through well-placed spies and collaborators. But the information they garnered could not replace qualified Soviet scientists, nor could it compensate for the perceived Soviet weakness in industry. Yet many Americans, both average civilians and political and military leaders, adopted this explanation, allowing them to blame the Soviet bomb on the treachery of some "red" scientists who had become traitors to their country. This narrative became even stronger when, in February 1950, the British theoretical physicist Klaus Fuchs confessed to spying for the Soviets. Fuchs had worked at Los Alamos and Oak Ridge during the war as a member of the British team; he had worked under Hans Bethe in the Theoretical Physics Division, specializing in the process of implosion so integral to the plutonium-style atomic bomb. After the war he continued to work for the British atomic energy project, and by the time he confessed to his crimes, he had had years of access to the most important secrets of both the United States and Great Britain.

Fuchs's arrest led investigators to his courier, Harry Gold, who had been working as a spy for the Soviets since the mid-1930s. Gold was apprehended in May 1950, and his interrogation led to the discovery of the Soviet spy ring that included Julius and Ethel Rosenberg and Ethel's brother, David Greenglass. The resulting media and political whirlwind, in which newspapers across the country ran daily stories about espionage almost continuously for the three years before the Rosenbergs were executed (even the *Bulletin of the Atomic Scientists* followed the scandal), meant it was easy for many Americans to accept the premise that Russia had "stolen" the bomb, and not developed atomic weapons on its own. FBI director J. Edgar Hoover wrote two articles that appeared in *Reader's Digest* in May 1951 and August 1952, respectively, titled "The Crime of the Century: The Case of the A-bomb Spies" and "Red Spy Masters in America." In the former he argued that, between Harry Gold's network and Klaus Fuchs, "the basic secrets of nuclear fission had been stolen."[9]

A more quantitatively based (yet no less specious) line of reasoning involved the ability of the Soviet Union to acquire fissionable material for use in atomic bombs. Most policymakers had taken Groves's and the CIA's analysis of the paucity of Soviet high-grade uranium at face value, but after August 1949, they were forced to account for their faulty estimation. One explanation was that the Soviets were able to acquire uranium from North Korea and some of its other satellite nations, places where American surveyors had not searched for uranium ore, and therefore possible locations for Soviet exploitation. Another story argued that the

United States itself had sent high-grade uranium to the Soviet Union. A *New York Times* article published on December 6, 1949, reported that a former air force major, George Racy Jordan, had testified under oath to the House Un-American Activities Committee (HUAC) that uranium and atomic information had been sent to the Soviet Union in 1943 and 1944 with the aid of Harry Hopkins, a close adviser to President Roosevelt. The story continued by implicating Henry Wallace, vice president under Roosevelt at the time, in overruling Leslie Groves to allow the shipments to occur. Jordan testified that he had seen suitcases full of uranium and atomic documents marked "Oak Ridge," together with letters on White House stationery signed "H. H." One of the letters, presumably from Hopkins, said that the writer "had a hell of a time getting this away from Groves."[10]

Overall, according to the *Times*, the testimony to HUAC revealed that at least two hundred pounds of uranium oxide, 220 pounds of uranium nitrate, an estimated twenty-five to forty pounds of uranium metal, and an undetermined number of barrels of heavy water were sent to the Soviet Union as part of the Hopkins exports. Other shipments by American companies included seven hundred pounds of uranium oxide and 220 pounds of uranium nitrate, sales that were made "with full knowledge and approval of the appropriate Government agencies." The orders were considered routine and not noteworthy at the time.[11] In the companies' defense, the strategic importance of this material was not widely known at the time, and uranium was used in many commercial applications, yet this story, and others like it, was enough to convince many Americans that the Soviets had used unwitting Americans to support their atomic program.

Another premise some Americans accepted (the fifth, if you are counting) was that the Soviets had somehow subverted the process of atomic bomb development by ignoring safety considerations and taking shortcuts the Americans would not take. Because the Soviets were devious, were devoid of God-fearing sensibilities, and operated under an "oriental mindset," they were willing to eschew many of the safety measures included in the U.S. process of building atomic bombs. In the October 1949 issue of the *Bulletin of the Atomic Scientists*, an issue dedicated to understanding how the Soviets built their bomb in such a short time, both Bernard Brodie of Yale University and Eugene Rabinowitch of the University of Illinois (cofounder of the *Bulletin*) embraced this reasoning. Brodie wrote that "many of the refinements introduced into the American processes to safeguard human life and capital equipment may have been dispensed with."[12] Rabinowitch argued that the Soviets had most likely used slave labor for the more dangerous tasks, and that this could have "considerably reduced the effort by eliminating the costly and extensive safety installations provided in all our facilities."[13]

The final explanation was that captured German scientists had done the heavy lifting in the Soviets' atomic bomb development. Of course, this ignored both the fact that the Germans were nowhere near building their own atomic bomb, and the fact that the United States and Great Britain had captured the best of the German scientists. Regardless, an article in the *New York Times*, published the day after Truman's announcement of the Soviet explosion and titled "German Scientists Held Aiding Soviet," reported that about two hundred German scientists had accepted jobs in the Soviet Union and had "contributed the know-how to the Russian industrial potential in the conclusion of the atomic explosions."[14] This story was credible to many Americans because, according to the historian Clarence Lasby, as early as 1948 the Republicans in Congress had publicly attacked the State Department and the administration for blocking the immigration of German specialists. The theory gained traction in the 1950s, both when critics of the army accused it of abandoning German scientists to the Soviet Union at the end of the Second World War, and then again after the launch of *Sputnik* in 1957. The idea that the Soviets had captured better Germans gave comfort to Americans who wanted to attribute Soviet scientific and technological achievements to something other than Communist capabilities. According to Lasby, "With no specific evidence to the contrary, millions of Americans accepted the thesis that the Truman administration had somehow been derelict in its importation program."[15]

Soviet Science

The detonation of the first Soviet atomic bomb on August 29, 1949, did not come as a surprise to all American atomic scientists. Many who had called for the internationalization of atomic energy had also warned against underestimating Soviet scientific capabilities. On June 11, 1945, a committee of Manhattan Project scientists working in the Metallurgical Laboratory of the University of Chicago wrote a memorandum to Secretary of War Henry Stimson. The Franck Committee, named after its chair, the Nobel laureate James Franck,[16] consisted of Franck, the physicist Donald Hughes, the radiation oncologist J. J. Nickson, the biophysicist Eugene Rabinowitch, the physicist J. C. Sterns,[17] the chemist Glenn Seaborg, and Leo Szilard. Their report warned Stimson that the Soviets most certainly had known the basic facts and implications of nuclear power as early as 1940, and that their scientists were sufficiently capable and experienced to "enable them to retrace [American] steps within a few years, even if [the United States made] all attempts to conceal them." At most, the Franck Report continued, it would take the Soviets only three or four years to construct their own atomic bomb, and even this assessment assumed that the U.S. program continued its own development. After eight to ten years, other nations, including the Soviet Union, could equal the United States in "intensive work in this field."[18]

In September 1945, three hundred Los Alamos scientists signed a memorandum about the future of atomic science and sent it through Robert Oppenheimer to George Harrison, professor of experimental physics and dean of science at MIT and head of the Office of Field Service of the OSRD during the Second World War. The Los Alamos scientists advised against using the American experience in building the atomic bomb as a guide to Soviet progress in atomic weapons (or that of any other nation). Although it took the United States six and a half years following the discovery of fission to build the bomb, the knowledge of the feasibility of the bomb—demonstrated at Hiroshima and Nagasaki—would compensate for the Americans' head start. As Glenn Seaborg would later write, "The only secret about the atomic bomb was whether or not it would work, and that question had been answered."[19] As a result of this knowledge, the Soviet Union could forgo much of the time-consuming laboratory experimentation the Americans had conducted, and according to the scientists, it was therefore "highly probable" that another nation such as the Soviet Union could join the United States as an atomic power "within a few years."[20] In November, the U.S. Senate Special Committee on Atomic Energy held hearings to formulate national policy on the development and control of atomic energy. One of the questions up for debate was the ability of the Soviet Union to match the accomplishment of the United States in atomic weapons. Irving Langmuir, a chemist and physicist who had won the 1932 Nobel Prize in Chemistry, told the committee that he believed the Soviets would have an atomic bomb in only three years.[21]

In 1946, American atomic scientists continued to warn the government and the American public against complacency. Early that year, the physicist and geochemist Harrison Brown, who had worked on the Manhattan Project and was the first to isolate larger quantities of plutonium for use in atomic bombs, wrote a book describing the dangers of atomic weapons, titled *Must Destruction Be Our Destiny?* In it, Brown argued that it was an "inescapable conclusion" that the Soviet Union would soon have its own atomic weapons. He pleaded for all Americans to "recognize that in another three years the United States of America may not stand alone as a possessor of atomic bombs."[22]

In February, an article in the *Bulletin of the Atomic Scientists* titled "Russia and the Atomic Bomb" detailed the "high level of research in nuclear physics in Russia." The article explained that the Soviet Union had enough scientific and industrial power to develop its own atomic bombs "within a few years," and that its scientists possessed the capability to engage in "extraordinarily skillful experimentation." While it acknowledged that the Soviet Union did not have an "array of great leaders" in atomic physics comparable to that of the United States, it argued that the Soviets, relying solely on their own scientific manpower, had enough highly qualified scientists to produce an atomic bomb in a short time.[23]

The opinions of these American scientists were informed by a deep understanding of the proficiency of Soviet science. Most had studied and worked with their Soviet counterparts in the scientific centers of Western Europe before the war. The Americans knew the Soviets were intelligent and capable, and that they came from a country with a long tradition of producing and supporting world-class scientists. Russia was, after all, the land of Ivan Pavlov, Leonhard Euler, and Dmitri Mendeleev. It was also the land of Abram Ioffe, a scientist who was considered the father of Russian atomic physics. At the turn of the century, Ioffe worked in Germany with the Nobel laureate Wilhelm Roentgen, the discoverer of X-rays. After earning his PhD at Munich University in 1905, Ioffe returned to Russia and founded the Institute of Physics and Technology (Fiztekh) in Saint Petersburg, where he supervised the instruction of an entire generation of Russian, and later Soviet, atomic scientists.[24]

Vladimir Vernadsky, a Russian mineralogist who had worked at the Curie Institute in Paris, realized as early as 1910 that radioactivity could lead to a new source of energy millions of times more powerful than anything then known. He founded the State Radium Institute in Petrograd (Saint Petersburg) in 1922 and continued to promote the development of atomic energy to the Soviet government throughout the interwar period. Yuly Khariton earned his doctorate in theoretical physics under Ernest Rutherford at the Cavendish Laboratory at Cambridge in 1927, Georgy Flerov discovered spontaneous fission in uranium in 1940,[25] and Lev Landau worked in Germany during the interwar period with the Manhattan Project physicist Edward Teller and the German-British physicist Rudolf Peierls.

The physicist Igor Kurchatov was also internationally known and highly respected. He was a protégé of Abram Ioffe, and in 1932 Ioffe appointed Kurchatov, then twenty-nine years old, to direct the nuclear physics program at Fiztekh. Kurchatov was a natural leader, and his enthusiasm, self-confidence, and abilities overshadowed the fact that he was quite young for such a prestigious position. As proof of his qualifications, in 1934 Kurchatov built Europe's first cyclotron, which was at the time the only operational cyclotron outside Berkeley, California.

Finally, the Soviet Union was the land of the physicist Peter Kapitza, who had also worked at the Cavendish Laboratory under Ernest Rutherford. He came to Cambridge in 1921 and remained there for over ten years, researching cryogenics and strong magnetic fields, and he became a Fellow of the Royal Society in 1929. In the 1930s, Kapitza formed the Institute for Physical Problems of the Russian Academy of Sciences in Moscow using equipment provided by Rutherford (Lev Landau was its head of the Theoretical Division). After the war, he was instrumental in the creation of the premier scientific laboratory in the Soviet Union, the Moscow Institute of Physics and Technology. Kapitza was closely acquainted with many Western scientists, including Niels Bohr, Robert Oppenheimer, Albert Einstein, Otto Hahn, and Victor Weisskopf, all of whom considered him to be among the world's elite. Weisskopf, in his memoirs, called Kapitza "one of the world's foremost experimental physicists."[26]

Despite the pleading of American scientists and the available evidence of the quality of Soviet personnel, however, the Americans' high opinion of science in the Soviet Union was far from universal. In general, the scientists who were warning the government against complacency had alienated themselves from those in power by advocating for the internationalization of atomic energy. Their voices were superseded by those who held a much lower opinion of Soviet science, and who promoted the maintenance of an American atomic monopoly. The American physicist Herbert York, who worked for the Manhattan Project at the Berkeley Radiation Laboratory and at Oak Ridge, wrote a memoir in 1978 in which he detailed the views of U.S. government scientists and policymakers toward the Soviet Union's scientific capabilities. York worked as a scientist for the government almost continuously from the Second World War through the 1980s. After the war, he continued his government nuclear work as the first director of the Lawrence Livermore National Laboratory in California, and later would be appointed chief scientist for the Defense Advanced Research Projects Agency and the director of defense research and engineering (now called assistant secretary of defense for research and engineering). In between his government work, York served as a professor of physics at Berkeley and chancellor of the University of California, San Diego, before ending his government service as the U.S. ambassador to the Comprehensive Test Ban negotiations in 1979–81. Arguably, more than anyone else, York had a unique perspective on the official and semiofficial perceptions of Soviet science during the Cold War.

In his memoir, York concedes that the lack of information about the state of Soviet science prohibited him, and the government, from making any concrete evaluation of Soviet capabilities. As a result of stringent security, Soviet progress in the sciences had been concealed from the rest of the world. According to York, the only Russian innovation Westerners were aware of was the two-seated farm tractor, "whose main function seemed to be to replace the church social as a place where Red Pioneer boys could meet collective-farm girls." Even "true-blue" European and American Communists did not think of the Soviet Union as a scientifically or technologically progressive nation. To American government scientists, the intelligence community, and the U.S. policymaking elite, the Soviet Union was "as mysterious and remote as the other side of the moon and not much more productive when it came to really new ideas or inventions." York concludes his evaluation of the Americans' perception of Soviet science with what he calls "a common joke of the time": the United States had time before it had to worry about the Soviet Union surreptitiously bringing atomic bombs in suitcases to destroy major U.S. cities, because the Soviets would first have to develop the technology of a suitcase.[27]

Maj. Gen. John Medaris provided an even less charitable view of Soviet science. Medaris was the commander of the U.S. Army Ballistic Missile Agency in the 1950s, and directed the German rocket scientist Wernher von Braun in the creation of the U.S. missile and satellite program. In his memoir, *Countdown for Decision*, Medaris argues that up until the launch of *Sputnik* in 1957, it was fashionable to think of Soviet scientists as "retarded folk who depended mainly on a few captured German scientists for their achievements, if any." Since the United States had captured the best and brightest German scientists during the war, "there was nothing to worry about."[28]

The result of these views, in part, was an intelligence community that predicted an eight-year grace period before the Soviets could produce their own atomic weapon. President Truman refused to respect even this conservative estimate. He was convinced that the Soviets, or "those Asiatics," as he called them,[29] would never match the scientific accomplishments of the United States and build their own atomic bomb. Leslie Groves told Congress at the end of the war that he thought it would take the Soviets at least fifteen to twenty years (more likely the latter) before they could replicate what the Americans had done.[30]

Groves's estimate was based less on the capabilities of Soviet scientists, and more on his belief in the inability of the Soviet Union to provide Soviet science with the industrial support necessary to build an atomic bomb. His assessment of Soviet industrial weakness was formed by three interdependent factors. First, he understood better than anyone what the United States needed to do in order to provide the Manhattan Project with the means to build its atomic arsenal. Second, he knew that German industry during the war, considered by many to be among the strongest in the world, could not provide the necessary support for its atomic program. Finally, he was told by representatives of U.S. industry that the Soviet Union did not have the available infrastructure to provide for such a massive undertaking as an atomic bomb project.

Groves understood that the key to the development of a successful nuclear weapon was the ability to translate theoretical physics abstractions and concepts into a tangible technological product. This meant creating and stockpiling adequate stocks of uranium and plutonium, both to provide for the immense amount of experimentation necessary in any atomic program and to provide the fuel for the bombs themselves. For the United States, this process had taken huge amounts of electric power, much of which was provided by the rural electrification projects of the New Deal in the 1930s, such as the Tennessee Valley Authority. Without the significant resources the United States put into industrial modernization before the war, the Manhattan Project would have lacked the sufficient infrastructural foundation to build the U.S. atomic bomb.

As early as May 1945, Groves had begun to speak to the leadership of U.S. industry to gauge Soviet industrial capabilities.[31] On May 21, Groves called G. M. Read of the DuPont Company and asked him how long it would take the Soviets to build a facility like the Hanford, Washington, installation that provided the Manhattan Project with fissile material. Read told Groves that the experience of DuPont in building an ammonia plant in Moscow demonstrated that Soviet skilled labor and mechanics "were so poor that they allowed machinery to pound itself to death." They also had difficulty finding capable men to run the plant once it was constructed, and Read believed that the Soviets would not have enough men to build, or run, an effective uranium separation facility. Even if they were given the exact plans and blueprints for the American plant, it would take the Soviet Union so long to reproduce what the Americans had accomplished that Read argued that the Soviets would not "live long enough to build one of these things."[32]

On June 1, Groves attended an Interim Committee meeting in which prominent American industrialists were invited to speak on this issue. The industrialists included George Bucher, the president of Westinghouse, whose company had manufactured the equipment for the electromagnetic process of isotope separation; Walter Carpenter, the president of DuPont; James Rafferty, the vice president of Union Carbide, the company that manufactured and operated the gas diffusion plant at Oak Ridge; and James White, the president of Tennessee Eastman, the producer of chemicals for the Manhattan Project and the company that constructed the RDX explosives plant in Tennessee. Secretary of War Stimson asked each of these men how long it would take for the Soviets to reproduce what their companies had built, based on their knowledge of their own efforts and their understanding of the Soviet Union's capabilities.

Walter Carpenter told the committee it had taken DuPont twenty-seven months to complete its Hanford facility, once they had received the basic plans. He explained that DuPont had to work with ten thousand to fifteen thousand other companies in order to complete Hanford on time, and without such help it would have taken significantly longer. Carpenter estimated that it would take the Soviet Union "at least four or five years" to construct this type of facility—that is, four or five years just to build the separation plant, not a bomb—and this assumed they already had the basic plans for the plant. He believed that the Soviets' greatest difficulty would be in securing the necessary skilled labor and technicians and adequate production facilities. James White stressed the advantage the United States had over the Soviet Union in standardized mass production capabilities. Special ceramics, vacuum tubes, special stainless steels, and "a great variety of special products" were needed in his Tennessee plant, and he doubted whether the Soviet Union "would be able to secure sufficient precision in its equipment to make this operation possible." He also echoed Carpenter's assertion that the Soviets would have difficulty finding enough skilled and educated personnel to catch the Americans.[33]

George Bucher of Westinghouse estimated that if the Soviets were able to utilize captured German technicians and scientists then they might be able to produce an electromagnetic pilot plant in as little as nine months, but it would take at least three years before the plant would be fully operational. He pointed out that the major problem the Soviets would need to overcome was that this type of plant required large numbers of replacement parts and "extremely accurate precision tools." James Rafferty of Union Carbide told the committee about the process employed at the gas diffusion plant at Oak Ridge. Metal uranium was converted to a gas and then U-235 was separated from U-238 by means of "extremely delicate barriers or screens." The barriers are the key to the entire process, and the Oak Ridge plant used over five million of them to produce fissionable material for atomic bombs. Rafferty estimated that the Soviets would require at least ten years to build this kind of plant without the basic knowledge only the Americans possessed. It would take the Soviets five years to develop just the barrier itself. According to Rafferty, the biggest problem the Soviets faced was a fundamental lack of experimental engineers, and even if the Soviet Union was given all of the necessary information about the plant's manufacture and the secrets of the barrier through espionage or other means, it would still take the Soviets a minimum of three years to get a gas diffusion plant into operation.[34]

As each industrialist testified, Groves became more and more confident of his estimation that it would take the Soviet Union more than a decade to build an atomic bomb. Three years later, in a June 1948 issue of the *Saturday Evening Post*, Groves explained his rationale to the American public. He wrote that the Soviet Union would be incapable of building an atomic bomb in less than a decade even if the United States had sent the "complete blueprints of the Manhattan Project to Russia on V-J Day." He emphasized the extent of the industrial effort, noting that the gaseous diffusion plant at Oak Ridge required twelve thousand construction drawings, fifteen thousand piping-material-erection sheets, and fifty thousand material-order sheets for its operation. Blueprints for the rest of the Oak Ridge project would cover approximately five hundred acres if they were spread out on the ground, and combined with the gaseous diffusion plans, they would weigh more than 230 tons. This does not even include all the plans that would be necessary for Soviet duplication of Hanford or Los Alamos, or the tens of thousands of special-design drawings made by U.S. industrial firms across the country. Groves concluded, "Once all these plans were collected, translated into Russian language and measurements, and safely delivered to the Soviet Union's top scientists, what would they do with them? If past experience is any criterion, they would waste a couple of years searching suspiciously for a gimmick in the plans, which, they would be confident, some American had fiendishly inserted to assure Russia the privilege of blowing herself off the map."[35]

Groves and many others in the military, intelligence, and policymaking communities were certain that the Soviets did not have the industrial capability to produce the facilities to develop an atomic bomb. The *Saturday Evening Post* article publicized this line of argument, but it was certainly not the first, and would not be the last, publication to do so. Over the period from 1945 to the summer of 1949, dozens of articles were written advocating this position in *Time* magazine, *Life* magazine, *Fortune*, the *Bulletin of the Atomic Scientists*, *Scientific American*, the *New York Times* (and the *New York Times Magazine*), the *Washington Post*, and regional newspapers throughout the United States. One of the most widely read and well-respected journalists of the day was the *New York Times* reporter Hanson Baldwin. As the newspaper's military editor (he wrote for the *Times* for forty years), Baldwin won the Pulitzer Prize for his reporting from Guadalcanal in 1943, and during his career authored eighteen books on military operations. His close association with the U.S. defense community allowed him to write an insider's-point-of-view article on the U.S. government's perception of the Soviets' atomic bomb prospects, which was published in the *Times* on November 9, 1947. The article, titled "Has Russia the Atomic Bomb?—Probably Not," and subtitled "Best American Opinion Is That She Will Need Years to Develop It," argued that although the Soviet Union, like most other countries, had the theoretical scientific knowledge necessary to build the bomb, it did not have the industrial capability, technical know-how, or manpower availability to build a bomb within a few years.[36]

Baldwin's article detailed the difficulties in the design and manufacture of the thousands of new and intricate devices—gauges, valves, instruments, piping, electrical devices—involved in the production of an atomic bomb. Baldwin contended that the manufacture of the bomb required engineers, technicians, administrative and production experts, machine tools, facilities, and general production knowledge "which Russia *very definitely* does not have in quality or quantity comparable to our own" (emphasis added). Despite its large population, he wrote, the Soviet Union could not concentrate an unlimited amount of energy or manpower on the production and development of atomic weapons unless it was to "neglect dangerously other major developments." Even if it were to do so, Baldwin argued, because of the "relatively low productivity of the Russian worker," the limited amount of industrial strength and electrical power available in the Soviet Union, and the scarcity of machine tools and skilled workers, it was unlikely that the Soviets could concentrate as much total energy on the production and development of atomic energy as the United States did during the Second World War. In addition, Baldwin was "reasonably certain" that the Soviet Union did not have manufacturing facilities comparable to Hanford, Oak Ridge, or Los Alamos. By 1947, he said, the Soviets had probably built "one or more simple atomic piles, but there is a long way from such a pile to the finished bomb."[37]

Baldwin concluded his article with a reassuring message to the American public: the United States had a "great headstart in the atomic race," and neither the Soviet Union nor any other country was likely to catch up in the foreseeable future. He reiterated that this was not his opinion, but instead the collective view of "responsible Government authorities" who had made a reassessment of Soviet atomic potential, resulting in "a dramatic change in attitude toward the short-term future." In 1945, atomic scientists had been "talking glibly" of ten thousand atomic bombs and had assured the public repeatedly that the Soviets would catch up to and overtake the United States in just a few years. In 1947, according to Baldwin, the government knew a different reality, one that meant "that some of the terrific sense of urgency that overhung all atomic bomb discussions two years ago has been removed; we still have time." He added, "Whether this is a benefit and will permit more mature, more reasoned and less passionate and hasty decisions, or whether elimination of the sense of urgency will induce complacency, only the future can tell."[38]

The perceived scarcity of fissile material was yet another reason for the underestimation of the Soviets' ability to manufacture an atomic bomb. From information it had gathered before the war, U.S. intelligence concluded that the Soviet Union did not have large deposits of high-quality uranium inside its borders.[39] Leslie Groves and other high-level American policymakers assumed that the Soviets would have a difficult time obtaining the necessary ore for atomic bomb development. While the Soviets did have control of the Joachimsthal mines in Czechoslovakia and the rebuilt Auergesellschaft Plant in Oranienburg, Groves's experience and his knowledge of the German program convinced him that this would not be enough to fulfill their materials requirements. In some ways Groves and other U.S. officials were correct: the Soviet Union had no domestic sources of high-grade uranium and would have to make do with low-grade ore with a uranium content of as little as 1–2 percent. What he and the others did not understand is that low-grade ore, which was found in abundance in the Soviet Union and almost everywhere else on earth, is entirely sufficient to begin the uranium refinement process. It might take longer to refine the uranium to weapons grade, but it could be done.

Groves did not fully understand the science behind uranium refinement, and as a result he took steps to prevent the Soviet Union from obtaining high-grade ore from foreign sources. He redirected a wartime policy originally intended to prevent the Germans from acquiring uranium ore and targeted it against the Soviet Union. The Combined Development Trust, an agreement between President Roosevelt and Prime Minister Churchill on June 13, 1944, gave Groves and the British Tube Alloys program a mandate to take control of all known available sources of uranium worldwide.[40] The trust gave Groves the ability to work outside the normal bureaucratic and government channels to aggressively pursue a monopoly (in theory) on fissionable material, and the Alsos Mission and the U.S. military gave him the geographical and geological information he needed to plan his acquisitions.[41] At the end of the war, the Combined Development Trust was designed to remain in place until it was extended or revised by official agreement. Groves used this provision to continue to control sources of uranium outside the United States, both to feed the accelerating U.S. nuclear weapons program and to deny these essential resources to the Soviets.[42]

In addition, Groves convinced the government to stop all shipments of equipment that could, in any conceivable way, be used in uranium production. In April 1946, the United States used the Coordinating Committee for Multilateral Export Control to further prohibit uranium-production items from going to the Soviet Union. The committee was created immediately after the Second World War, and its membership included the countries that would become NATO (as well as Japan). It was designed to keep strategic materials out of the hands of the Soviets and their allies, and in this case Groves used it to prevent equipment such as vacuum pumps, high-temperature heat-resistant steel (called sicromal), and other essential equipment from reaching the Soviet atomic bomb program. According to Henry Lowenhaupt, a scientist who worked in Groves's Foreign Intelligence Section of the MED, and who would later serve a long career in the CIA, "Export control pressure against the Russian atomic program was being applied as rapidly and as forcefully as we could arrange it."[43]

Convinced that the Combined Development Trust was keeping uranium away from the Soviets, and hopeful that export controls could prevent them from acquiring the necessary equipment for uranium refinement, Groves was confident that he had found the way to impede Soviet progress toward the development of an atomic bomb.

The Soviet System and the Atomic Bomb

The perceived incompatibility of the totalitarian Soviet system with advanced scientific discovery and innovative technological development was another component in the Americans' underestimation of the Soviet capability to produce atomic weapons. The Soviet Union demanded universal acceptance of Marxist-Leninist ideology from its scientists, and refusal to adhere to the dogma meant the end of a career, banishment to a forced labor camp, or even death. The result was a system in which politics and science were inseparable.[44]

All of this was widely known to many Americans in the 1940s, and certainly to most American scientists. Waldemar Kaempffert, the science editor for the *New York Times*, wrote an article in September 1946 illustrating the history and effects of Marxism-Leninism on Soviet science. "Science—and Ideology—in Soviet Russia" detailed the implementation of Soviet political philosophy in science following the Russian Revolution. Once the Communist Party consolidated its political power, it moved to purge the Academy of Sciences of suspected dissenters and counterrevolutionaries. Many scientists were dismissed or imprisoned, and the ones who remained scrambled to profess their faith in the Soviet system by publishing articles with titles such as "Marxism and Surgery," "The Dialectics of Graded Steel," and "The Dialectics of the Internal Combustion Engine."[45]

The most widely publicized influence of political ideology on science was in Soviet biology. Following the revolution in 1917, Russian biologists tried to convince the Soviet leadership that entire species could be transformed through changed environmental conditions. As the species adapted to struggle, they would progress, and become a better version of what they were before. As a logical extension of Marxism, the scientists argued, this theory must be correct. Yet this premise is antithetical to Gregor Mendel's science of genetics, which was accepted as valid by scientists worldwide and, most importantly, by Vladimir Lenin, who threw his full support behind the Soviet biologist Nikolai Vavilov, the most prominent Russian geneticist. Vavilov would become a member of the Soviet Academy of Sciences, president of the Lenin Academy of Agricultural Sciences, and director of the Institute of Applied Botany. He was a foreign member of the Royal Society of London and was considered for membership as a foreign associate in the U.S. National Academy of Sciences.[46] Yet none of this could protect him from the changes that would occur in the Soviet Union.

After Lenin's death and the purges of "dissenters" and "counterrevolutionaries" in the late 1920s and 1930s, the scientists who opposed the genetic theory of biology were elevated to positions of prominence by the Soviet political hierarchy—no one more so than the botanist Trofim Lysenko, who argued that the theory of genetics was inconsistent with Marxist philosophy. Lysenko accused Vavilov of introducing to the Soviet Union foreign scientific ideas that came from fascist Germany and capitalist Great Britain and the United States. Lysenko's science, on the other hand, was a *Soviet* science, and the fact that no other nation's scientists subscribed to his theories only proved that the Soviet system had provided the impetus for the next true advancement in biology. By 1940, Lysenko had convinced the Soviet leadership to replace Vavilov with himself as director of the Genetics Institute of the Academy of Sciences and the Institute of Applied Botany. Vavilov was exiled to a forced labor camp in Siberia, where he died in 1942.[47]

As the Second World War came to a close, and as Western scientists began to learn about the rise of Lysenko and the death of Vavilov, dozens of articles began to appear in American scientific journals, and even mainstream periodicals, about the subservience of science to social and political philosophy in the Soviet Union. The Harvard biologist Vladimir Asmous published a particularly scathing report on Lysenkoism in the March 1946 issue of *Science*, in which he condemned the Soviet system for subjugating science to politics and argued that "freedom, as Americans understand it, is simply nonexistent in [the] USSR." He continued, "But the most disturbing fact is that the case of Vavilov is by no means an exception. We know that hundreds of less-known Russian scientists are dying slowly in Soviet concentration camps which can compete quite favorably in atrocities with Belsen, Dachau, and other Nazi horror camps."[48]

Most of the articles published in the United States about Lysenkoism were written by biologists, geneticists, and botanists, but it would not be long before the U.S. physics community began to link the ideology of Lysenko to Soviet physics. An article in the December 1948 issue of the *Bulletin of the Atomic Scientists* warned that the attack on "bourgeois" influence in Soviet science had been "extended to the field of atomic physics" when the Soviet Union accused four of its physicists of subscribing to the "reactionary idealism and formalism" of Niels Bohr's Copenhagen school of nuclear physics.[49] The trend was so alarming to American atomic physicists that the *Bulletin* dedicated its entire May 1949 issue to Lysenkoism and how it affected Soviet atomic physics.

The editors of the *Bulletin* explained that while it might seem strange that the journal would devote an issue to a review of events in the Soviet Union in the field of genetics, "the Soviet purge of genetics is of deep concern to scientists everywhere and to the *Bulletin* in particular, because it is an extreme expression of a development in the opposite direction—toward even greater disregard of scientific facts and methods, and subordination of science to political expediency." They continued,

> The supremacy of a racial, social, or economic dogma over the whole spiritual life of a nation, claimed by the modern totalitarian states, has created a new and ominous threat. It is not merely that the pursuit of science has been declared to derive its only justification from immediate benefits to society. . . . What is novel and alarming is that science is not only restricted, but also perverted.

The dogma of Lysenko, the editors argued, was being ruthlessly imposed on the entire scientific community of almost half the world, and genetics was only the beginning of the problem:

> What we lament is not merely the brutal destruction of a flourishing branch of science, the interruption of the life work of a number of good scientists, the wrecking of their laboratories, and the uncertainty of their personal fate. The supreme misfortune is the reversion of a large part of Europe to prescientific dogmatism, at a time when the survival of our civilization requires universal readiness to abide by scientifically established facts, and to use objective scientific methods in dealing with the crucial problems of mankind—problems such as atomic energy control, the prevention of war, and the rational utilization of world resources.

The editors concluded with a warning to Western scientists:

> Another lesson of the purge worth pondering by American scientists is that science cannot remain permanently unfettered in a system which exercises strict control over other activities of the human mind—religion, philosophy, literature, art, social and economic research. For a long time, science has appeared as a happy island of free thought in the sea of Soviet regimentation. Not only was it supported on a scale which was the envy of many Western scientists, but except for occasional incursions, it was left free to pursue its self-set aims according to its own rules.

> We state here these lessons of the purge, not as our contribution to the "cold war," but to encourage a long-range perspective as to the consequences of "statism" for the free growth of science.[50]

In the April 1949 issue of *Philosophy of Science*, an article by Lewis Feuer detailed the negative effects of Marxist political philosophy on Soviet physics. Feuer, a sociologist and former dedicated Marxist, was a professor at Vassar College and would later write one of the most widely read books on Marxist theory.[51] Feuer argued that Soviet philosophy had prevented the country's physicists from participating in the great advances in physical science. Beginning with its criticism of the theory of relativity as idealistic and metaphysical, the Soviet government discriminated against the followers of Einstein for twenty-five years, starting at the time of the revolution. Finally, after more than two decades of debate, Soviet scientists were able to integrate relativity with Marxism. By this time, however, the Soviet system, according to Feuer, had significantly "retarded the development of Soviet physical science." Western scientists, unencumbered by dogma, were able to continue their work throughout the 1920s and 1930s. Soviet scientists spent twenty-five years trying to catch up to their Western counterparts, until they could finally invent phrases like "the dialectical unity of time and space" that would allow them to operate within the same scientific framework as the rest of the world.[52]

Lysenkoism also criticized the use of probability and statistical methods in science. According to Lysenko, "All the so-called laws of Mendel and Morgan are built on the ideas of accident—but genuine science is the enemy of accident."[53] Quantum physics, the branch of physics most heavily utilized in the theory and development of atomic weapons, relies on probability and statistical methods, and thus under Lysenkoism it would not be regarded as "genuine science." For Feuer, this helped to explain why the Soviet Union was behind the United States in atomic development: "Soviet physicists who bear the baggage of their philosophic doctrine are impeded in their work. They must be mindful that their methods conform not only to the facts but to the [Marxist] ideology; the two conditions cannot both be always satisfied. Perhaps the failure of Soviet physics to achieve the Western successes in atomic theory and invention are partially due to the wasteful influence of the philosophy of [Marxism]."[54]

The critique of Soviet science was not limited to academics in universities. Many key American policymakers, or those who influenced them, were also highly critical of the impact of political philosophy on Soviet science. Leslie Groves certainly was. Groves, whose avid anti-Communism dated back to before the Second World War, thought the Soviet system would slow the progress of Soviet atomic energy development. In November 1945, while testifying before the Senate Special Committee on Atomic Energy, Groves responded to critics in Congress who argued that his estimate of the time when the Soviet Union would build its first atomic bomb was wrong. Groves admitted that he may have misestimated the Soviet timeline, but he insisted that if so, it might be "an error in the other direction." Instead of fifteen or twenty years, it could be forty to fifty. Based on conversations he had had with associates who had visited the Soviet Union, Groves was told that, because of the Soviet system, the Soviet Union might never develop atomic weapons. The rationale for this conclusion was that the Soviets, under their present system, would never get "men with courage enough to go in and make the mistakes that are necessary to produce such a thing as this."[55]

Samuel Goudsmit, the scientific chief of the Alsos Mission, joined the debate in 1947. After the war, Goudsmit continued his work in atomic energy and became a senior scientist at the Brookhaven National Laboratory in New York. In his book *Alsos*, Goudsmit wrote about the failure of the Germans to develop atomic weapons, and attributed that failure to the inability of science to function in a totalitarian system. To make his case, Goudsmit used the same terminology as the critics of Lysenkoism had been using, and although there is no direct reference to the Soviet Union in the book, it is unlikely that this was coincidental:

> German science, as we have seen, was severely handicapped by Nazi dogma. By persecuting and exiling all scholars afflicted with the Jewish "taint," Germany lost some of the greatest scientists in the world. In a healthy country, however, such a loss could have been replaced in a relatively short time by outstanding scholars who were followers of the exiled men. This did not happen in Germany because the effect of the Nazi ideology was to make "non-Aryan" sciences like modern physics, unpopular, with a consequent loss of promising students. Finally, the instruction of the few students who dared to study the abstract, or "non-Aryan" sciences, progressively

> deteriorated. Quite frequently the Nazis appointed teachers who did not even understand what they were teaching. Thus Munich, under the great Sommerfeld, was once the world's most productive university in theoretical physics. When Sommerfeld retired, shortly before the war, he was replaced by a Nazi named Muller, who did not "believe" in modern physics.[56]

Goudsmit argued that the German experience could provide important lessons for scientists and policymakers in the postwar world:

> Too many of us still assume that totalitarianism gets things done where democracy only fumbles along, and that certainly in those branches of science contributing directly to the war effort the Nazis were able to cut all corners and proceed with ruthless and matchless efficiency. Nothing could be further from the truth. . . . The failure of German physics can in large measure be attributed to the totalitarian climate in which it lived. There are lessons we can all learn from that failure. . . . Politics, the interference of politicians in the affairs of science, and the appointment of party hacks to important administrative posts, is another grave error it would be foolish to suppose was purely a German monopoly. . . . The same thing applies to dogma, whether it be political, scientific or religious. The stubborn blindness of dogma and the free inquiring spirit of science do not mix.[57]

The most important and influential voice for science in the postwar United States was Vannevar Bush. As the director of the Office of Scientific Research and Development during the war, Bush worked closely with members of the scientific community, the military, the intelligence community, and the political hierarchy. After the war, Bush was appointed director of the Joint Research and Development Board of the Army and Navy (which became the Research and Development Board of the Department of Defense after 1947), and his influence at all levels of science and government continued. Like Groves and Goudsmit, Bush believed that the Soviet system of government would prohibit the Soviets from achieving significant scientific innovations or technological developments.

Bush had held this view since, at the latest, 1945. In an Interim Committee meeting on May 31 of that year, Bush told the membership that the United States' advantage over "totalitarian states" during the war had been "tremendous." Evidence from Germany had demonstrated that the American advantage "stemmed in large measure from [the American] system of teamwork and free interchange of information by which [the United States] had won out and would continue to win out in any competitive scientific and technological race."[58]

In 1949, Bush published *Modern Arms and Free Men*, in which he explained his views on the Soviet system of totalitarian government and the Soviets' capability to match the United States in atomic energy and in general scientific development. Bush argued that the Soviet system's rigidity meant that the United States had years before it would have to worry about a Soviet nuclear power:

> It has also been grasped that the task of repeating what this country did under the pressure of war is no mean task and requires years of effort. Thus the time has been moved ahead when there may be two stocks of bombs of comparable and substantial size, and we have more breathing time than we once thought. There is a high probability that there are some years, perhaps quite a few, before the question of two prospective belligerents frowning at each other over great piles of atomic bombs can become a reality. . . . The time estimate depends, of course, on how fully we think our adversaries may put their backs into the effort, how much they are willing, or able, to reduce their standard of living in order to accomplish it. They lack men of special skills, plants adapted to making special projects, and possibly materials. As we shall discuss later, they lack the resourcefulness of free men, and regimentation is ill adapted to unconventional efforts. On the other hand, their tight dictatorship can order effort, no matter how much it hurts. But we do not need an exact estimate; it is sufficient to note that opinion now indicates a longer time than it did just after the end of the war. The problem is not altered in its nature by this more moderate estimate; it is certainly less critical and immediate.[59]

Bush contended that the weakness of the Soviet Union was its ideological rigidity. It could not tolerate diversity, and this was "fatal" to true progress in fundamental science. A dictatorship, like that in the Soviet Union, could not tolerate independence of thought and expression, and commitment to the party line prevented science from flourishing under such a system. Regardless of individual genius, he argued, a great scientist cannot operate in a system in which he is sent into exile if he questions the official position of the state, no matter how antithetical to science it might be. The development of a Soviet atomic bomb could be the most affected by this type of rigidity:

> The keynote of all this effort [the Manhattan Project] was that it was on an essentially democratic basis, in spite of the necessary and at times absurd restrictions of secrecy and the formality that tends to freeze any military, or for that matter governmental, operation of great magnitude. If certain physicists thought the organization was functioning badly in certain respects, they could walk in on the civilian who headed that aspect of the effort and tell him so in no uncertain terms. They not only could, they most certainly did; and the point is that there was no rancor, and old friendships were not destroyed in the process. If civilians and military disagreed, as they often did, there were tables about which they could gather and argue it out. Punches did not need to be pulled, and no one kept glancing over his shoulder. If there were international misunderstandings between allies, and there were, they could be frankly discussed, sometimes with more heat than light, but also with a prevailing atmosphere of genuine desire to arrive at the conclusion that made sense and that best got on with the war. If a young scientist had an idea he did not have to pass it through a dozen formal echelons and wait a year; he could talk it over with his fellows and with superiors of accepted eminence in his own field and be sure it would be weighted with unbiased judgment by men of competence. The system worked and it produced results.[60]

The Nazis, on the other hand, were regimented in a totalitarian system. Their able physicists should have made better progress than they did, but the German totalitarian organization, Bush said, "was an abortion and a caricature." The German military leadership who commanded the program—or, as Bush called them, "nincompoops with chests full of medals"—presided over scientific operations of which they knew nothing. Thus communications between the scientists and the military were lacking, and the system prevented real innovation. According to Bush, this same type of system was present in the Soviet Union:

> The type of pyramidal totalitarian regime that the Communists have centered in Moscow is an exceedingly powerful agency for cold war. It is capable of holding great masses of people in subjection, indoctrinating them in its tenets, and marshalling them against the free world. It can force its

> people to enormous sacrifice and thus build great quantities of materials of war. It can educate large numbers of men and women in science and engineering, construct far-flung institutes, mechanize agriculture, and ultimately create mass production of the manifold things it needs. But it is not adapted for effective performance in pioneering fields, either in basic science or in involved and novel applications. It has many of the faults of the German dictatorship, magnified to the *n*th degree. Hence it is likely to produce great mistakes and great abortions.[61]

Bush concluded his argument with a reassuring message to the American people: the Soviet system of government could not possibly advance science with full effectiveness. It could not even apply science to war as effectively as the United States. Moreover, until the Soviet Union changed its system and became a free nation, it would not be able to alter its pattern of inefficiency or become fully successful, and if it did become free, "the contest is ended."[62]

CONCLUSION

Credit Where Credit Is Due

Nuclear weapons were the result of the direct application of cutting-edge advanced science to weapons development. The development of the atomic bomb during the Second World War was thus a transcendent moment in the history of the relationship between government and science. The atomic bomb revealed the practical potential of basic or theoretical science, or what could emerge when theory is transformed directly into functional technology. Furthermore, it demonstrated what could be accomplished when government supported large-scale scientific research and development with significant resources. The combination of these factors established a new pattern that changed not only traditional relations between science and government, but also how we think about applied science. This pattern would continue to develop throughout the second half of the twentieth century.

As a consequence, intelligence about an enemy nation's scientific capabilities became an essential component of strategic planning. Scientific intelligence, however, was unlike any type of intelligence that had come before it. Instead of focusing on tangible threats or existing materials, scientific intelligence was designed to predict the potential future ramifications of scientific research and development. To do so, scientific intelligence professionals had to evaluate the general capabilities of a nation, and then determine whether it had the ability to develop atomic weapons. They then used those particular assessments to determine the proximity and magnitude of the prospective strategic danger. It was not an ideal process, but because of the military impact of nuclear weapons it became an absolute necessity.

Why was the U.S. government unable to create an effective atomic intelligence apparatus to monitor Soviet scientific and nuclear capabilities? How did we get this so wrong?

In the Second World War, the U.S. government assumed that the advanced nature of German science favored the success of a serious endeavor to build atomic weapons. As a result, the U.S. leadership expended immense resources and effort to determine the extent of the German atomic bomb program. In 1942 the U.S. scientific, intelligence, military, and political leadership faced the unprecedented challenge of creating a scientific intelligence system capable of assessing the extent of foreign atomic development. This action was precipitated by an acute fear of German capabilities, a fear that originated in an extremely high regard for German science and the belief that if the Germans produced an atomic weapon, they would not hesitate to use it against the Allies. Because the fear of German abilities was so pronounced, the U.S. government allowed Leslie Groves to create a strong, centralized, and coordinated system of atomic intelligence. In turn, this organization was extraordinarily effective, and capable of providing actionable intelligence that successfully challenged the presuppositions of U.S. leadership concerning the German atomic bomb program.

The centralization of U.S. scientific and atomic intelligence was the key component in the success of the American effort against the German atomic bomb program. When the first U.S. scientific intelligence program was initially conceived by scientists, the effort yielded poor results. The American scientists who had taken it upon themselves to learn all they could about the German atomic program were not skilled in intelligence collection or analysis, and thus were unqualified for the task at hand. At the same time, intelligence professionals in the United States did not have the scientific knowledge to inform their efforts, and were equally as unsuccessful as the scientists in learning the extent of the German atomic program. In addition, the established intelligence agencies—the army's G-2, the Office of Naval Intelligence, the Office of Strategic Services, and a number of smaller intelligence agencies within governmental organizations such as the State Department—did not coordinate their atomic intelligence efforts, resulting in significant gaps in intelligence coverage. And since they were not under a single, integrated command, parochialism and bureaucratic infighting prevented the established intelligence agencies from operating at an effective level.

The solution to this problem was to consolidate the atomic intelligence program under Brig. Gen. Leslie Groves, who was well versed in the scientific and technological fields that atomic theory encompassed, and who also had a general knowledge of the intricacies of intelligence collection, analysis, and dissemination. His background in large-scale construction and engineering had trained him to manage complicated tasks and disparate groups, and he had the full confidence of George Marshall, Vannevar Bush, and President Roosevelt.

Leslie Groves will be remembered for his successful direction of the Manhattan Project, and rightfully so. But he should also be given credit as the man most responsible for the creation of the first centralized U.S. scientific intelligence organization. When he took command of atomic intelligence in the fall of 1943, Groves immediately set out to consolidate all atomic intelligence functions under his individual control. He acquired the cooperation of G-2, the ONI, and the OSS, securing their promise to send all atomic information his way. To handle the day-to-day intelligence operations, he appointed trusted subordinates such as John Lansdale, Robert Furman, Tony Calvert, and Boris Pash, each of whom was not only highly competent but also fiercely loyal to Groves. He exploited British intelligence sources, bringing the entirety of the information gathered by the combined Allied atomic intelligence effort under his control. Finally, through overt military and clandestine operations, Groves was able to take actions that retarded German progress toward the development of an atomic bomb.

Groves's centralized and integrated atomic intelligence organization was immensely successful, and met or exceeded expectations at all three levels of the intelligence cycle. The paragon of the collection effort was, of course, the Alsos Mission, which accomplished all of its goals and should be considered one of the most successful intelligence operations in history. But the collection efforts of the MED intelligence team were not limited to Alsos. Lansdale, Furman, and Calvert gathered information from a wide variety of sources, compiling as complete a picture of the German atomic program as was possible and, in doing so, providing the Alsos Mission with all the resources it would need to be successful.

Timely analysis of German atomic intelligence was equally effective in Groves's centralized system. Highly qualified scientists, such as James Fisk and Samuel Goudsmit, were able to perform on-site analysis of German atomic developments while deployed with Alsos. In addition, the MED intelligence team utilized its own cohort of world-class scientists in the United States. As information was gathered, either by the Alsos Mission or by the collection efforts of Lansdale, Furman, and Calvert, the assets of MED intelligence were able to build an accurate assessment of German atomic development. By the time the Alsos Mission reached Strasbourg, Groves's centralized organization had laid the analytical groundwork to prepare MED intelligence for the dramatic revelation that Germany was years away from building an atomic bomb.

Just as important as the MED intelligence team's effective collection and analysis was its ability to convince American policymakers of the true state of German atomic development, and to disabuse the U.S. leadership of its perception that German scientific and atomic research would naturally outpace that of the United States. Both Vannevar Bush and Leslie Groves trusted the soldiers and scientists of Alsos enough to immediately accept the Strasbourg evidence as valid. The trust they in turn had garnered with the political and military leadership ensured that the status of the German atomic bomb program would be accepted by top policymakers such as Marshall and Roosevelt.

However, the Soviet atomic energy program was perceived by many Americans as incapable of accomplishing the task of building an atomic bomb within a few years of the end of the war. The United States regarded Soviet science as substandard, and it was presupposed that the inferiority of Soviet science would prevent the Soviet Union from producing an atomic bomb before the mid-1950s. In addition, Americans widely believed that the Soviets did not possess the industrial capabilities to develop atomic weapons in any quantity in such a short time period.

Some also argued that the rigidity of the Soviet totalitarian system would prevent the Soviet Union from quickly manufacturing an atomic bomb. Regardless of the reasoning, these assumptions gave the U.S. scientific, military, and political leadership the misguided impression that they had ample time before the Soviet Union could catch up with the United States in atomic development, and thus the maintenance of a strong, centralized atomic intelligence program was not an immediate high priority.

Despite its wide-ranging success and years of compiled institutional knowledge and experience, the U.S. centralized atomic intelligence system was dismantled after the Second World War, its personnel and resources strewn throughout the various remaining intelligence agencies of the U.S. government. While many of the personnel continued to work on atomic intelligence issues, they did so in a decentralized, disjointed manner that was not capable of providing policymakers with an accurate picture of Soviet scientific development in the atomic field. The resulting atomic intelligence organization failed in all three aspects of the intelligence cycle. Collection was done piecemeal, through a variety of intelligence organizations, and could not provide analysts with the information necessary to produce an accurate assessment of the Soviet atomic program. Without adequate raw data, analysts made estimates that were based mainly on wild speculation about what they assumed the Soviet Union would and could do. In many cases, these estimates were based solely on the American and German experiences, and not in any way on actual information from the Soviet Union. As a result, both military and civilian policymakers were given the impression that the Soviet atomic program was not of immediate concern, and that they could continue to pay it—and the improvement of the U.S. atomic intelligence system—less attention than it ultimately deserved. The poor performance of U.S. atomic intelligence meant that the faulty estimates of the Soviet nuclear program would continue, thereby slowing any measures to improve the U.S. atomic intelligence system.

Even the detonation of the first Soviet atomic bomb in August 1949 did not convince most Americans to reconsider their perception of Soviet science. In the immediate aftermath of the detection of the explosion, the U.S. scientific, military, and policymaking elite spread blame widely for the intelligence failure, but refused to acknowledge the possibility of

Soviet scientific strength as the primary culprit. Instead they latched on to ideas that mitigated the impact of Soviet scientific ability.

While the rest of the U.S. national security system was improving (primarily as a result of the provisions of national legislation such as the National Security Act of 1947 and NSC-68), the refusal to give Soviet science credit where credit was due meant that the U.S. scientific intelligence apparatus continued to falter well into the 1950s. The CIA's Office of Scientific Intelligence (OSI)—which was explicitly created to collect, analyze, and disseminate intelligence concerning enemy scientific development—did not become an effective intelligence agency until the 1960s, despite the emerging Soviet atomic threat.

An OSI survey report released in early 1952 indicated that the U.S. scientific intelligence effort against the Soviet Union was experiencing considerable problems. It still lacked the necessary scientifically trained personnel, and their need for scientific intelligence, the report observed, went "far beyond" their capacity to collect raw intelligence "susceptible of accurate evaluation." The basic deficiency in the OSI's scientific intelligence product, according to the report, stemmed from the "abysmal gaps" in the Americans' knowledge of the state of research in the Soviet Union. The survey placed the bulk of the blame for this failure on a lack of centralization. This dynamic continued to exist not only within the CIA, but also between the CIA and the military services, where rivalries persisted over the question of whose "exclusive prerogatives" scientific and atomic intelligence fell into. These interservice and interagency rivalries resulted in fragmented collection, disjointed analysis, information not being shared within the scientific intelligence community, and ultimately a continued failure of the system.[1]

This would not begin to change until after 1957 and the launch of *Sputnik*. Again the United States was surprised by the scientific abilities of the Soviet Union. This time, however, it was forced to accept the realization that the Soviet Union had equaled, and in some cases even surpassed, the scientific capabilities of the United States. Once the U.S. intelligence community reached this conclusion, the rebuilding of an effective scientific intelligence apparatus could begin. In 1963, the CIA formed the Directorate of Science and Technology, and over the next decade consolidated all of the CIA's scientific and intelligence functions under its auspices.

The detonation of the first Soviet atomic bomb would, however, alter the type of intelligence that pertained to the Soviet atomic bomb program. Since the Soviet Union had completed its development of atomic weapons, and they were no longer in the research stage, the Soviet atomic program stopped being the purview of scientific intelligence and became technological intelligence. Yet scientific intelligence directed at the Soviet Union would remain consequential even after 1949. Other potential Soviet scientific developments, such as nuclear weapons delivery systems (e.g., long-range missiles and submarines), anti–ballistic missile systems, scientific implications for conventional weapons systems, and improvements in biological and chemical weapons, would continue to be relevant to U.S. national security policymakers.

Yet there it seems the U.S. national security community would, in the end, resign itself to the fact that they might never learn enough about any Soviet scientific research for defense purposes. Instead the U.S. political and military leadership decided to rely on a security policy that included offensive military buildup (both nuclear and conventional), defensive military buildup (missile defense, a strong fighter-interceptor program, attack submarines for Soviet missile submarine interdiction), mutual deterrence, and technological intelligence (the U-2 and SR-71 programs, reconnaissance satellites).[2] It wasn't a perfect solution. But since we are still here to argue about it?

Good enough.

Notes

Introduction

1. During the Second World War, U.S. atomic intelligence made no considerable effort to collect information on atomic developments in Japan. There were several reasons Japan was dismissed as a potential atomic threat. First, it was believed that Japan did not have the necessary raw materials to produce an atomic weapon. Second, U.S. intelligence assumed that Japan did not have the necessary industrial capability for an atomic program on the scale needed to produce deployable atomic weapons. Third, while the American scientific community greatly respected their Japanese counterparts, the scientists told the intelligence community that the qualified and capable Japanese atomic scientists were too few in number to allow Japan to produce an atomic bomb. Finally, unlike the situation in Germany, the Japanese had given U.S. intelligence no indications that they were interested in building an atomic bomb.

1. A Reasonable Fear

1. There were certainly other scientists involved, but while it might be unfair to their contributions to exclude them here, Rutherford and Soddy were the principal scientists who should be credited with this discovery. Rutherford would receive the Nobel Prize in Chemistry in 1908 for his work at McGill.

2. Frederick Soddy, *Radio-Activity: An Elementary Treatise* (London: "The Electrician," 1904), 170.

3. Ernest Lawrence's cyclotron accelerated protons to thousands of times the speed of sound and at millions of volts of energy.

4. Quoted in Richard Rhodes, *The Making of the Atomic Bomb* (New York: Simon and Schuster, 1986), 209.

5. From 1924 to 1934, Szilard filed for twenty-nine patents from the German patent office.

6. Rhodes, *Making of the Atomic Bomb*, 203.

7. Leo Szilard, *The Collected Works: Scientific Papers* (Cambridge, MA: MIT Press, 1972), 642; Rhodes, *Making of the Atomic Bomb*, 214.

8. Rhodes, *Making of the Atomic Bomb*, 212.

9. The number 2.5×10^{21} is reached by dividing Avogadro's number by 238. Avogadro's number tells us how many atoms are in a mole of an element. There are 238 (or 235) moles in a single gram of uranium, thus the above calculation.

10. Rhodes, *Making of the Atomic Bomb*, 274–75.

11. Einstein to Roosevelt, August 2, 1939, in *The American Atom: A Documentary History of Nuclear Policies from the Discovery of Fission to the Present*, ed. Philip Cantelon, Richard Hewlett, and Robert Williams (Philadelphia: University of Pennsylvania Press, 1984), 9–11.

12. Leslie R. Groves, *Now It Can Be Told: The Story of the Manhattan Project* (New York: Da Capo, 1975), 7.

13. Richard G. Hewlett and Oscar E. Anderson Jr., *A History of the United States Atomic Energy Commission*, vol. 1, *The New World, 1939–1946* (Berkeley: University of California Press, 1990), 42–43.

14. Amir D. Aczel, *Uranium Wars: The Scientific Rivalry That Created the Nuclear Age* (New York: Palgrave Macmillan, 2009), 132–33.

15. Arthur Holly Compton, *Atomic Quest: A Personal Narrative* (New York: Oxford University Press, 1956), 223.

16. Aczel, *Uranium Wars*, 161.

17. Werner Karl Heisenberg, interview conducted and edited by Joseph J. Ermenc, August 29, 1967, in *Atomic Bomb Scientists: Memoirs, 1939–1945*, ed. Joseph J. Ermenc (Westport, CT: Meckler, 1989), 55.

18. Arthur Compton to Vannevar Bush, June 22, 1942, RG 227, Bush-Conant Files, M1392, roll 7, National Archives and Records Administration II, College Park, MD (hereafter cited as NARA II).

19. Compton, *Atomic Quest*, 118, 221.

20. Siegfried Flügge, "Kann der Energieinhalt der Atomkerne Technisch Nutzbar Gemacht Werden?," *Die Naturwissenschaften* 27, nos. 23/24 (June 1939): 402–10; Thomas Powers, *Heisenberg's War: The Secret History of the German Bomb* (New York: Knopf, 1993), 10; Jeffrey T. Richelson, *Spying on the Bomb: American Nuclear Intelligence from Nazi Germany to Iran and North Korea* (New York: W. W. Norton, 2006), 20.

21. Harold Urey to Vannevar Bush, June 26, 1942, RG 227, Bush-Conant Files, M1392, roll 7, NARA II.

22. Leo Szilard to Arthur Compton, June 1, 1942, RG 227, Bush-Conant Files, M1392, roll 7, NARA II.

23. Rhodes, *Making of the Atomic Bomb*, 118.

24. Groves, *Now It Can Be Told*, 33.

25. David C. Cassidy, *Beyond Uncertainty: Heisenberg, Quantum Physics, and the Bomb* (New York: Bellevue Literary Press, 2009), 306.

26. Compton, *Atomic Quest*, 87–88.

27. Compton, *Atomic Quest*, 221.

28. Samuel A. Goudsmit, *Alsos*, vol. 1 of *The History of Modern Physics, 1800–1950* (New York: Henry Schuman, 1947), 32.

29. Goudsmit, *Alsos*, 113.

30. Aczel, *Uranium Wars*, 116–17.

31. Matrix mechanics explained, among other things, how quantum jumps occur within an atom (electrons "jumping" from one energy level to another). It was an extension of the Bohr model of the atom.

32. American Institute of Physics, "Werner Heisenberg," accessed June 9, 2015, http://www.aip.org/history/heisenberg/p08.htm.

33. Victor Weisskopf, *The Joy of Insight: Passions of a Physicist* (New York: Basic Books, 1991), 31.

34. Aczel, *Uranium Wars*, 131–32; Powers, *Heisenberg's War*, 5–8; Weisskopf, *Joy of Insight*, 130.

35. Glenn T. Seaborg, *Adventures in the Atomic Age: From Watts to Washington* (New York: Farrar, Straus and Giroux, 2001), 57.

36. In 1909 Haber had developed a method for extracting nitrogen from the air to artificially make ammonia, predominantly for use as a fertilizer. Later, this process would become strategically essential for the Germans' production of nitrates for explosives. Germany had no nitrates of its own; instead it had counted on the importation of sodium nitrates from the northern desert of Chile. During the First World War, the Allies successfully cut off this supply from the Germans. Without Haber's discovery, the Germans would not have been able to last nearly as long as they did.

37. Aczel, *Uranium Wars*, 67.

38. Aczel, *Uranium Wars*, 68–69; Arnold Kramish, *The Griffin* (Boston: Houghton Mifflin, 1986), 185–86.

39. Paul Harteck, interview conducted and edited by Joseph J. Ermenc, July 6, 1967, in Ermenc, *Atomic Bomb Scientists*, 89–90.

40. Kramish, *Griffin*, 153.

41. The Stern-Gerlach experiment of 1922 tested the deflection of particles and is often used in universities today to demonstrate laws of angular momentum and some of the basic properties of quantum mechanics. It can be used to demonstrate that atomic particles (mainly electrons) have quantum properties.

42. Aczel, *Uranium Wars*, 138.

43. Einstein to Roosevelt, August 2, 1939.

44. Urey to Bush, June 26, 1942.

45. Kramish, *Griffin*, 244.

46. Goudsmit, *Alsos*, xxvii.

47. Groves, *Now It Can Be Told*, 6.

48. It seems this was a correct assertion. Albert Speer, the German minister of armaments and munitions, director of German war production, and a close friend and deputy of Hitler, wrote in his memoir *Inside the Third Reich* (New York: MacMillan, 1970),

> I am sure that Hitler would not have hesitated for a moment to employ atom bombs against England. I remember his reaction to the final scene of a newsreel on the bombing of Warsaw in the autumn of 1939. We were sitting with him and Goebbels in his Berlin salon watching the film. Clouds of smoke darkened the sky; dive bombers tilted and hurtled toward their goal; we could watch the flight of the released bombs, the pull-out of the planes and the cloud from the explosions expanding gigantically. The effect was enhanced by running the film in slow motion. Hitler was fascinated. The film ended

with a montage showing a plane diving toward the outlines of the British Isles. A burst of flame followed, and the island flew into the air in tatters. Hitler's enthusiasm was unbounded. 'That is what will happen to them!' he cried out, carried away. 'That is how we will annihilate them!" (227).

49. Powers, *Heisenberg's War*, 354.

50. Arthur Compton to James Conant, July 15, 1942, RG 227, Bush-Conant Files, M1392, roll 7, NARA II.

51. Groves, *Now It Can Be Told*, 199.

52. Compton, *Atomic Quest*, 222; Groves, *Now It Can Be Told*, 203.

53. Goudsmit, *Alsos*, 7–8.

54. Hewlett and Anderson, *History*, 60–61; Rhodes, *Making of the Atomic Bomb*, 405–6.

55. Compton, *Atomic Quest*, 102.

56. Goudsmit, *Alsos*, 7.

57. Szilard to Compton, June 1, 1942.

58. Seaborg, *Adventures*, 82.

59. Compton to Bush, June 22, 1942.

2. Making Something out of Nothing

1. Leslie R. Groves, *Now It Can Be Told: The Story of the Manhattan Project* (New York: Da Capo, 1975), 10; Arnold Kramish, *The Griffin* (Boston: Houghton Mifflin, 1986), 128; Victor Weisskopf, *The Joy of Insight: Passions of a Physicist* (New York: Basic Books, 1991), 119–20.

2. Allen Dulles, in his book *The Craft of Intelligence* (Guilford, CT: Lyons, 1963), provides information about technological intelligence during the American Revolution, the Civil War, and the Spanish-American War.

3. Reports of the Captured Enemy Material Units are found in RG 165, Records of the War Department General and Special Staffs, Office of the Director of Intelligence, G-2, Washington Liaison Branch, Security-Classified General Correspondence, 1943–1945, entry 185, box 124, National Archives and Records Administration II, College Park, MD (hereafter cited as NARA II).

4. Arthur Compton to James Conant, July 15, 1942, RG 227, Bush-Conant Files, M1392, roll 7, NARA II.

5. Arthur Compton to J. C. Sterns, July 16, 1942, RG 227, Bush-Conant Files, M1392, roll 7, NARA II.

6. Jeffrey T. Richelson, *Spying on the Bomb: American Nuclear Intelligence from Nazi Germany to Iran and North Korea* (New York: W. W. Norton, 2006), 27–29; Thomas Powers, *Heisenberg's War: The Secret History of the German Bomb* (New York: Knopf, 1993), 167–68; Arthur Compton, *Atomic Quest: A Personal Narrative* (New York: Oxford University Press, 1956), 221–24.

7. Leo Szilard to Arthur Compton, June 1, 1942, RG 227, Bush-Conant Files, M1392, roll 7, NARA II.

8. Harold Urey to Vannevar Bush, June 26, 1942, RG 227, Bush-Conant Files, M1392, roll 7, NARA II.

9. The Wigner memorandum of June 20, 1942, is still classified. This is most likely the case because it contains key information about the production of plutonium that is still relevant today. However, over the last seventy years, information from a number of sources (fellow scientists, military personnel, and Wigner himself) has revealed the tenor of Wigner's memorandum.

10. Vannevar Bush to James Conant, June 30, 1942, RG 227, Bush-Conant Files, M1392, roll 7, NARA II.

11. Vannevar Bush to Maj. Gen. George Strong, July 6, 1942, RG 227, Bush-Conant Files, M1392, roll 7, NARA II.

12. Vannevar Bush to Maj. Gen. George Strong, September 21, 1942, RG 227, Bush-Conant Files, M1392, roll 7, NARA II.

13. Memorandum attached to Bush to Strong, September 21, 1942.

14. Memorandum attached to Bush to Strong, September 21, 1942.

15. Powers, *Heisenberg's War*, 192.

16. Powers, *Heisenberg's War*, 190–94; David C. Cassidy, *Beyond Uncertainty: Heisenberg, Quantum Physics, and the Bomb* (New York: Bellevue Literary Press, 2009), 360; Weisskopf, *Joy of Insight*, 119.

17. Weisskopf, *Joy of Insight*, 119.

18. "Present Status and Future Program," report by Vannevar Bush to the Military Policy Committee, December 15, 1942, Correspondence ("Top Secret") of the Manhattan Engineer District, 1942–1946, RG 77, M1109, roll 3, NARA II.

19. Harry Wensel to James Conant, September 30, 1942, RG 227, Bush-Conant Files, M1392, roll 7, NARA II.

20. Groves, *Now It Can Be Told*, 185–86.

21. Vannevar Bush, *Modern Arms and Free Men: A Discussion of the Role of Science in Preserving Democracy* (New York: Simon and Schuster, 1949), 135.

22. Samuel A. Goudsmit, *Alsos*, vol. 1 of *The History of Modern Physics, 1800–1950* (New York: Henry Schuman, 1947), 10–11. Interestingly, both Bush and Goudsmit wrote their books at about the same time, so unless they spoke during the writing process, the "Mata Hari" references are original to both authors.

23. Groves, *Now It Can Be Told*, 185.

24. The infighting between G-2 and the OSS became so contentious that officers in G-2 more than once reported OSS agents to the FBI for security violations.

25. Groves, *Now It Can Be Told*, 185–86.

26. Groves, *Now It Can Be Told*, 185.

27. Groves, *Now It Can Be Told*, 186.

28. Groves, *Now It Can Be Told*, 190.

29. Leslie R. Groves, "The A-Bomb Program," in *Science, Technology, and Management*, ed. Fremont Kast and James Rosenzweig (New York: McGraw-Hill, 1963), 40.

30. Groves, "A-Bomb Program," 32.

31. Until this time, atomic research was nominally being directed by the OSRD, Vannevar Bush, and James Conant (Bush's deputy and the president of Harvard University).

32. The idea was to have representatives from the army, navy, and OSRD. Vannevar Bush was the chairman and James Conant the alternate chair; from the army was Maj. Gen. W. D. Styer (chief of staff to General Somervell, commanding general of Army Service Forces), and from the navy was Rear Adm. W. R. E. Purnell (assistant to Admiral King, the chief of naval operations). While Groves was officially outranked only by the military members of the committee, in the broader governmental hierarchy, both Bush and Conant were his superiors. Groves, "A-Bomb Program," 33.

33. Groves, "A-Bomb Program," 33.

34. Groves, "A-Bomb Program," 33.

35. Groves, "A-Bomb Program," 34.

36. Groves, "A-Bomb Program," 37.

37. Groves, "A-Bomb Program," 35.

38. Groves, "A-Bomb Program," 40.

39. Groves, "A-Bomb Program," 39. Just before Hiroshima his office would double to ten rooms, but only because Groves understood that he needed to add a public relations staff in order to deal with the aftermath of the atomic bombings of Japan.

40. Groves, "A-Bomb Program," 40.

41. Groves, *Now It Can Be Told*, xii–xiii.

42. Groves, *Now It Can Be Told*, 186.

43. John Lansdale Jr., "Military Service" (unpublished manuscript, 1987), Wood Library–Museum of Anesthesiology, 82, http://woodlibrarymuseum.org/ebooks/item/157/lansdale-john-jr-military-service.

44. Lansdale, "Military Service," 1–3.

45. Lansdale, "Military Service," 11.

46. Lansdale, "Military Service," 23–24.

47. Lansdale, "Military Service," 7.

48. Robert S. Norris, *Racing for the Bomb: General Leslie R. Groves, the Manhattan Project's Indispensable Man* (South Royalton, VT: Steerforth, 2002), 10, 198; Dennis Hevesi, "R. R. Furman, 93, Dies; Led Bomb-Project Spying," *New York Times*, October 30, 2008; C. Peter Chen, "Robert Furman," World War II Database, accessed November 15, 2008, http://ww2db.com/person_bio.php?person_id=476.

49. John Keegan, *Intelligence in War: Knowledge of the Enemy from Napoleon to Al-Qaeda* (New York: Alfred A. Knopf, 2003), 259–60.

50. Goudsmit, *Alsos*, 9.

51. Groves, *Now It Can Be Told*, 187.

52. Powers, *Heisenberg's War*, 223; Richelson, *Spying on the Bomb*, 32.

53. Groves, *Now It Can Be Told*, 199–200.

54. Philip Morrison to Samuel Allison, September 23, 1943, RG 77, entry 22, box 170, NARA II; Samuel Allison to Brig. Gen. Leslie Groves, October 11, 1943, RG 77, entry 22, box 170, NARA II.

55. Richelson, *Spying on the Bomb*, 35.

56. Robert Oppenheimer to Maj. Robert Furman, September 22, 1943, box 34, J. Robert Oppenheimer Papers, Manuscript Division, Library of Congress.

57. Richelson, *Spying on the Bomb*, 34.

58. Powers, *Heisenberg's War*, 228; Richelson, *Spying on the Bomb*, 37.

59. Groves, *Now It Can Be Told*, 195.

60. Groves, *Now It Can Be Told*, 196.

61. Groves, *Now It Can Be Told*, 197–98.

62. Groves, *Now It Can Be Told*, 198.

63. Robert Jungk, *Brighter Than a Thousand Suns: A Personal History of the Atomic Scientists* (New York: Harcourt, Brace, 1958); David Irving, *The German Atomic Bomb* (New York: Simon and Schuster, 1967); Powers, *Heisenberg's War*; Elisabeth Heisenberg, *Inner Exile: Recollections of a Life with Werner Heisenberg* (Boston: Birkhauser, 1984), 76–80. A more nuanced and balanced discussion of the Bohr-Heisenberg meeting can be found in Richard Rhodes, *The Making of the Atomic Bomb* (New York: Simon and Schuster, 1986), 383–86.

64. Jungk, *Brighter Than a Thousand Suns*, 99.

65. Jungk, *Brighter Than a Thousand Suns*, 101.

66. Werner Heisenberg to Robert Jungk, 1958, in Jungk, *Brighter Than a Thousand Suns*, 103.

67. Jungk, *Brighter Than a Thousand Suns*, 104.

68. Aage Bohr, "The War Years and the Prospects Raised by the Atomic Weapons," in *Niels Bohr: His Life and Work as Seen by His Friends and Colleagues*, ed. Stefan Rozental (Amsterdam: North-Holland, 1967), 193.

69. Niels Bohr to Werner Heisenberg, undated (written sometime after 1957), document 1, Documents Relating to 1941 Bohr-Heisenberg Meeting, Niels Bohr Archive, https://www.nbarchive.dk/collections/bohr-heisenberg/.

70. Niels Bohr to Werner Heisenberg, document 7, Niels Bohr Archive.

71. Lansdale, "Military Service," 47.

72. Groves, *Now It Can Be Told*, 196.

73. The MAUD Committee consisted of its chair, Sir George Paget Thomson (physicist, Nobel laureate, and son of the Nobel laureate J. J. Thomson); the physicist Marcus Oliphant; P. M. S. Blackett, physicist and father of Operations Research; James Chadwick, discoverer of the neutron; the experimental physicist Philip Moon; and the physicist John Cockcroft.

74. Kramish, *Griffin*, 104.

75. Powers, *Heisenberg's War*, 195.

76. R. V. Jones, *The Wizard War: British Scientific Intelligence, 1939–1945* (New York: Coward, McCann, and Geoghegan, 1978), 205–6, 472–75.

77. Jones, *Wizard War*, 477–78.

78. For more on this, see Powers, *Heisenberg's War*, 282–85; and F. H. Hinsley, *British Intelligence and the Second World War: Its Influence on Strategy and Operations*, vol. 3, part 2, 584–86.

79. Robert Furman to Leslie Groves, February 12, 1946, Correspondence ("Top Secret") of the Manhattan Engineer District, 1942–1946, RG 77, M1109, roll 2, NARA II. For more information on the Norwegian raid on Norsk Hydro, see Dan Kurzman, *Blood and Water: Sabotaging Hitler's Bomb* (New York: Henry Holt, 1997); Thomas Gallagher, *Assault in Norway: Sabotaging the Nazi Nuclear Bomb* (New York: Bantam Books, 1981); and Jones, *Wizard War*.

80. War Cabinet Offices to Joint Staff Mission, Washington, April 6, 1943, Correspondence ("Top Secret") of the Manhattan Engineer District, 1942–1946, RG 77, M1109, roll 2, NARA II.

81. War Cabinet Offices to Joint Staff Mission, Washington, April 7, 1943, Correspondence ("Top Secret") of the Manhattan Engineer District, 1942–1946, RG 77, M1109, roll 2, NARA II.

82. "Paraphrase of Telegram Just Received from a Reliable Source," n.d., Correspondence ("Top Secret") of the Manhattan Engineer District, 1942–1946, RG 77, M1109, roll 2, NARA II.

83. Rhodes, *Making of the Atomic Bomb*, 512.

84. Maj. Gen. George Strong to Gen. George Marshall, August 13, 1943, Correspondence ("Top Secret") of the Manhattan Engineer District, 1942–1946, RG 77, M1109, roll 2, NARA II.

85. Rhodes, *Making of the Atomic Bomb*, 512–14.

86. Rhodes, *Making of the Atomic Bomb*, 517.

87. Strong to Marshall, August 13, 1943.

88. "Nazi 'Heavy Water' Looms as Weapon," *New York Times*, April 4, 1943.

89. Urey wrote a number of letters to the science editors of most major American periodicals, including *Time* magazine, the *New Republic*, and many others. Copies can be found in Correspondence ("Top Secret") of the Manhattan Engineer District, 1942–1946, RG 77, M1109, roll 2, NARA II.

90. Lansdale, "Military Service," 13–22.

91. Lansdale, "Military Service," 23–24.

92. According to Lansdale, "For example, we were required to discontinue the maintenance of centralized investigative records but these were essential to our work." Lansdale, "Military Service," 43–44.

93. Lansdale, "Military Service," 43–44; Groves, *Now It Can Be Told*, 139.

94. Lansdale, "Military Service," 24–25.

95. Lansdale, "Military Service," 25. Tom Clark later became the U.S. attorney general and, subsequently, an associate justice of the U.S. Supreme Court.

96. Lansdale, "Military Service," 27. J. Edgar Hoover was aware of the Manhattan Project, but very few others were informed.

97. Franklin Roosevelt's opinion of the State Department's ability to keep a secret: "That place is a sieve." Quoted in Meredith Hindley, *Destination Casablanca: Exile, Espionage, and the Battle for North Africa in World War II* (New York: Hachette, 2017), 194.

98. Lansdale, "Military Service," 47.

3. Alsos

1. John Lansdale Jr., "Military Service" (unpublished manuscript, 1987), Wood Library–Museum of Anesthesiology, 41–42, http://woodlibrarymuseum.org/ebooks/item/157/lansdale-john-jr-military-service.

2. Leslie R. Groves, *Now It Can Be Told: The Story of the Manhattan Project* (New York: Da Capo, 1975), 190.

3. George Strong to George Marshall, September 25, 1943, Correspondence ("Top Secret") of the Manhattan Engineer District, 1942–1946, RG 77, M1109, roll 4, National Archives and Records Administration II, College Park, MD (hereafter cited as NARA II).

4. Groves, *Now It Can Be Told*, 191.

5. Strong to Marshall, September 25, 1943.

6. Groves, *Now It Can Be Told*, 191–92.

7. Groves, *Now It Can Be Told*, 191.

8. Carroll Wilson to Vannevar Bush, "Scientific Mission to Italy," September 23, 1943, Correspondence ("Top Secret") of the Manhattan Engineer District, 1942–1946, RG 77, M1109, roll 4, NARA II.

9. Wilson to Bush, "Scientific Mission to Italy."

10. Wilson to Bush, "Scientific Mission to Italy."

11. Vannevar Bush to Leslie Groves, September 27, 1943, Correspondence ("Top Secret") of the Manhattan Engineer District, 1942–1946, RG 77, M1109, roll 4, NARA II.

12. Wilson to Bush, September, 23, 1943.

13. Bush to Groves, September, 27, 1943.

14. John Lansdale to George Strong, "Naval Participation in Alsos Detachment," November 4, 1943, Correspondence ("Top Secret") of the Manhattan Engineer District, 1942–1946, RG 77, M1109, roll 4, NARA II.

15. Frank Knox to Henry Stimson, November 10, 1943, Correspondence ("Top Secret") of the Manhattan Engineer District, 1942–1946, RG 77, M1109, roll 4, NARA II.

16. Henry Stimson to Frank Knox, "Proposed Mission to Investigate Secret Scientific Developments of the Enemy," November 16, 1943, Correspondence ("Top Secret") of the Manhattan Engineer District, 1942–1946, RG 77, M1109, roll 4, NARA II.

17. This is something NACA could not provide.

18. Lansdale, "Military Service," 28–29.

19. Groves, *Now It Can Be Told*, 192.

20. Lansdale, "Military Service," 29.

21. Boris Pash, *The Alsos Mission* (New York: Award House, 1969), 11.

22. John Johnson replaced George Kistiakowsky in early November. Robert Furman to Leslie Groves, November 10, 1943, Correspondence ("Top Secret") of the Manhattan Engineer District, 1942–1946, RG 77, M1109, roll 4, NARA II.

23. Robert Furman to Leslie Groves, October 20, 1943, Correspondence ("Top Secret") of the Manhattan Engineer District, 1942–1946, RG 77, M1109, roll 4, NARA II.

24. Pash, *Alsos Mission*, 15.

25. Groves, *Now It Can Be Told*, 192.

26. Henry Stimson to Dwight Eisenhower, November 26, 1943, Correspondence ("Top Secret") of the Manhattan Engineer District, 1942–1946, RG 77, M1109, roll 4, NARA II.

27. Pash, *Alsos Mission*, 15.

28. The official history of the mission sets the date for this as December 14, as do Groves's and Pash's personal recollections, but the date of the Progress Report that describes this event is December 13.

29. Boris Pash to John Lansdale, "Progress Report on Alsos (1)," December 13, 1943, Correspondence ("Top Secret") of the Manhattan Engineer District, 1942–1946, RG 77, M1109, roll 4, NARA II.

30. Allied Control Commissions managed the territory of defeated enemies.

31. Different documents have different spellings for Zaroli's name (Zarolli or Zirolli) and different ranks (commodore rather than captain).

32. Boris Pash to John Lansdale, handwritten progress report for Alsos (2), n.d., Correspondence ("Top Secret") of the Manhattan Engineer District, 1942–1946, RG 77, M1109, roll 4, NARA II.

33. Pash to Lansdale, "Report of Alsos Mission," December 30, 1943, Correspondence ("Top Secret") of the Manhattan Engineer District, 1942–1946, RG 77, M1109, roll 4, NARA II.

34. Pash to Lansdale, handwritten progress report.

35. Pash to Lansdale, "Report of Alsos Mission," December 30, 1943.

36. "Non-technical Report from Alsos Mission," January 20, 1944, Correspondence ("Top Secret") of the Manhattan Engineer District, 1942–1946, RG 77, M1109, roll 4, NARA II.

37. "Non-technical Report from Alsos Mission."

38. Pash to Lansdale, "Report of Alsos Mission."

39. Pash, *Alsos Mission*, 20.

40. Pash, *Alsos Mission*, 22.

41. Boris Pash to John Lansdale, January 21, 1944, Correspondence ("Top Secret") of the Manhattan Engineer District, 1942–1946, RG 77, M1109, roll 4, NARA II.

42. Pash to Lansdale, January 21, 1944.

43. Office of Strategic Services (OSS) Official Dispatch, Bern to OSS, Washington, DC, March 24, 1944, and May 11, 1944, RG 226, entry 134, box 219, NARA II.

44. OSS Official Dispatch, March 24, 1944, and May 11, 1944.

45. While with the Red Sox, he famously gave a young Ted Williams hitting tips. Williams, arguably the best hitter in the history of baseball, learned from Berg the intellectual foundations of the game, such as how to read pitchers and predict pitch selection, and what made greats such as Babe Ruth and Lou Gehrig so great.

46. The best single source on Moe Berg is Nicholas Dawidoff's *The Catcher Was a Spy* (New York: Vintage Books, 1994); also see Ralph Berger, "Moe Berg," Society for American Baseball Research (SABR) Baseball Biography Project, accessed June 9, 2009, http://sabr.org/bioproj/person/e1e65b3b.

47. His baseball card is on display at CIA headquarters.

48. Dawidoff, *Catcher Was a Spy*, 129–33; Berger, "Moe Berg"; Jewish Virtual Library, "Moe Berg," accessed June 9, 2009, http://www.jewishvirtuallibrary.org/jsource/biography/MBerg.html; "A Look Back . . . Moe Berg: Baseball Player, Linguist, Lawyer, Intel Officer,"

CIA Featured Story Archive, accessed June 9, 2009, https://www.cia.gov/news-information/featured-story-archive/2007-featured-story-archive/moe-berg.html.

49. Born's students included Robert Oppenheimer, Victor Weisskopf, and Pascual Jordan.

50. Vannevar Bush to Leslie Groves, February 1, 1944, Correspondence ("Top Secret") of the Manhattan Engineer District, 1942–1946, RG 77, M1109, roll 4, NARA II.

51. Pash, *Alsos Mission*, 29.

52. Will Allis, James Fisk, John Johnson, and Bruce Old, "Alsos Mission, Summary Report," January 22, 1944, Correspondence ("Top Secret") of the Manhattan Engineer District, 1942–1946, RG 77, M1109, roll 4, NARA II.

53. James Fisk to Leslie Groves, "(Interim) Report of Alsos Mission Together with Recommendations for the Obtaining of Scientific Intelligence," February 12, 1944, Correspondence ("Top Secret") of the Manhattan Engineer District, 1942–1946, RG 77, M1109, roll 4, NARA II.

54. James Fisk, "Interim Report from Alsos Mission," February 5, 1944, Correspondence ("Top Secret") of the Manhattan Engineer District, 1942–1946, RG 77, M1109, roll 4, NARA II.

55. Fisk, "Interim Report." In the February 14 report, Fisk wrote that Calosi had told Alsos that the Germans had "no 'astonishing' secret weapons." James Fisk, "Interim Report from Alsos Mission," February 14, 1944, Correspondence ("Top Secret") of the Manhattan Engineer District, 1942–1946, RG 77, M1109, roll 4, NARA II.

56. "Notes on Interview with Dr. Fisk," February 19, 1944, Correspondence ("Top Secret") of the Manhattan Engineer District, 1942–1946, RG 77, M1109, roll 4, NARA II.

57. "Notes on Interview with Dr. Fisk."

58. Lansdale, "Military Service," 38.

59. Lansdale, "Military Service," 41.

60. Groves, *Now It Can Be Told*, 194.

61. Vannevar Bush to Leslie Groves, February 29, 1944, Correspondence ("Top Secret") of the Manhattan Engineer District, 1942–1946, RG 77, M1109, roll 4, NARA II.

62. Carroll Wilson to Robert Furman, March 18, 1944, Correspondence ("Top Secret") of the Manhattan Engineer District, 1942–1946, RG 77, M1109, roll 4, NARA II.

63. Boris Pash to Leslie Groves, March 6, 1944, Correspondence ("Top Secret") of the Manhattan Engineer District, 1942–1946, RG 77, M1109, roll 4, NARA II; Robert Furman to Leslie Groves, March 6, 1944, Correspondence ("Top Secret") of the Manhattan Engineer District, 1942–1946, RG 77, M1109, roll 4, NARA II.

64. Furman to Groves, March 6, 1944.

65. Bissell had replaced Strong in January 1944.

66. Leslie Groves to Clayton Bissell, "Report of Alsos Mission," March 10, 1944, Correspondence ("Top Secret") of the Manhattan Engineer District, 1942–1946, RG 77, M1109, roll 4, NARA II.

67. This mainly applied to the supplementary personnel, such as CIC agents and interpreters who were borrowed from other units in the field.

68. Clayton Bissell to George Marshall, "Investigation of the Enemy's Secret Scientific Developments," April 1, 1944, Correspondence ("Top Secret") of the Manhattan Engineer District, 1942–1946, RG 77, M1109, roll 4, NARA II.

69. Leo James Mahoney, "A History of the War Department Scientific Intelligence Mission (ALSOS), 1943–1945" (PhD diss., Kent State University, 1981), 132, microfilm.

70. The deputy chief of staff signed the order in the name of the secretary of war.

71. Robert Furman to Leslie Groves, April 5, 1944, Correspondence ("Top Secret") of the Manhattan Engineer District, 1942–1946, RG 77, M1109, roll 4, NARA II.

72. John Weckerling to Leslie Groves, "Mission for Collecting Intelligence of the Enemy's Secret Scientific Developments," April 8, 1944, Correspondence ("Top Secret") of the Manhattan Engineer District, 1942–1946, RG 77, M1109, roll 4, NARA II.

73. Timeline of Alsos Mission created by Boris Pash and sent to Samuel Goudsmit, Samuel A. Goudsmit Papers, American Institute of Physics Collections, Niels Bohr Library & Archives (hereafter cited as AIP), http://www.aip.org/history/nbl/collections/goudsmit/index.html.

74. Mahoney, "History of the War Department," 142–43.

75. Robert Furman to Leslie Groves, "Status of New Alsos Mission," April 12, 1944, Correspondence ("Top Secret") of the Manhattan Engineer District, 1942–1946, RG 77, M1109, roll 4, NARA II.

76. Furman to Groves, April 12, 1944.

77. Furman to Groves, April 12, 1944.

78. "Boris Pash to Clayton Bissell, Attention C. P. Nicholas," April 12, 1944, Correspondence ("Top Secret") of the Manhattan Engineer District, 1942–1946, RG 77, M1109, roll 4, NARA II.

79. Groves, *Now It Can Be Told*, 208.

80. Clayton Bissell to George Marshall, "Mission Organized in MID for the Collection of Scientific Intelligence," May 11, 1944, Correspondence ("Top Secret") of the Manhattan Engineer District, 1942–1946, RG 77, M1109, roll 4, NARA II.

81. Alan Waterman to C. P. Nicholas, "SIM," June 10, 1944, Correspondence ("Top Secret") of the Manhattan Engineer District, 1942–1946, RG 77, M1109, roll 4, NARA II.

82. Samuel A. Goudsmit, *Alsos*, vol. 1 of *The History of Modern Physics, 1800–1950* (New York: Henry Schuman, 1947), 15.

83. Samuel Goudsmit to C. P. Nicholas, "Scientific Intelligence," May 15, 1944, Correspondence ("Top Secret") of the Manhattan Engineer District, 1942–1946, RG 77, M1109, roll 4, NARA II.

84. Pash, *Alsos Mission*, 13.

85. Samuel Goudsmit, "Final Report from the Scientific Chief of the Alsos Mission," December 7, 1945, Goudsmit Papers, AIP.

86. Pash, *Alsos Mission*, 38.

87. Boris Pash to Chief, Military Intelligence Service, June 13, 1945, Correspondence ("Top Secret") of the Manhattan Engineer District, 1942–1946, RG 77, M1109, roll 2, NARA II; Pash, *Alsos Mission*, 38.

88. Pash, *Alsos Mission*, 38.

89. Boris Pash to Chief, Military Intelligence Service, "Progress Report #1, Alsos Mission," July 24, 1944, Correspondence ("Top Secret") of the Manhattan Engineer District, 1942–1946, RG 77, M1109, roll 4, NARA II.

90. Pash, *Alsos Mission*, 31.

91. Pash, *Alsos Mission*, 32.

92. Morris Berg to Howard Dix, June 12, 1944, RG 226, entry 210, box 431, NARA II.

93. Groves, *Now It Can Be Told*, 209; W. M. Adams to Coordinator of Research and Development, Navy Department, June 7, 1944, Correspondence ("Top Secret") of the Manhattan Engineer District, 1942–1946, RG 77, M1109, roll 4, NARA II.

94. Pash to Chief, "Progress Report #1."

95. Boris Pash to Chief, Military Intelligence Service, "Progress Report #2, Alsos Mission," July 26, 1944, Correspondence ("Top Secret") of the Manhattan Engineer District, 1942–1946, RG 77, M1109, roll 4, NARA II.

96. Pash, *Alsos Mission*, 45.

97. Boris Pash to Leslie Groves, "Report on Plans for Paris Operation," August 15, 1944, Correspondence ("Top Secret") of the Manhattan Engineer District, 1942–1946, RG 77, M1109, roll 4, NARA II.

98. Boris Pash to Chief, Military Intelligence Service, "Progress Report—Alsos Mission—France No. 1," September 1, 1944, Correspondence ("Top Secret") of the Manhattan Engineer District, 1942–1946, RG 77, M1109, roll 4, NARA II.

99. Groves, *Now It Can Be Told*, 211.

100. Pash, *Alsos Mission*, 56.

101. The documents are not specific about the exact date.

102. Boris Pash to Chief, Military Intelligence Service, "Progress Report—Alsos Mission—France No. 2," September 7, 1944, Correspondence ("Top Secret") of the Manhattan Engineer District, 1942–1946, RG 77, M1109, roll 4, NARA II.

103. "Interview with Professor F. Joliot, London, September 5th and 7th, 1944," Office of the Chief of Engineers, S-1 History of Atomic Bomb, RG 77, entry A1 20, container 148, NARA II.

104. Goudsmit, *Alsos*, 34.

105. "Interview with Professor F. Joliot."

106. Groves, *Now It Can Be Told*, 215.

107. Groves, *Now It Can Be Told*, 201.

108. Boris Pash to Chief, Military Intelligence Service, "Progress Report No. 3—Alsos Mission, France," September 16, 1944, Correspondence ("Top Secret") of the Manhattan Engineer District, 1942–1946, RG 77, M1109, roll 4, NARA II.

109. Pash, *Alsos Mission*, 76.

110. Pash to Chief, "Progress Report No. 3."

111. Groves, *Now It Can Be Told*, 218.

112. Pash, *Alsos Mission*, 83.

113. Pash, *Alsos Mission*, 83.

114. Pash, *Alsos Mission*, 83.

115. Furman had just recently returned to Washington from Europe.

116. Boris Pash to Chief, Military Intelligence Service, "Progress Report No. 6 (Operations), Alsos Mission, France," October 6, 1944, Correspondence ("Top Secret") of the Manhattan Engineer District, 1942–1946, RG 77, M1109, roll 4, NARA II.

117. Boris Pash to Chief, Military Intelligence Service, "Progress Report No. 7 (Operations), Alsos Mission, France," October 12, 1944, Correspondence ("Top Secret") of the Manhattan Engineer District, 1942–1946, RG 77, M1109, roll 4, NARA II.

118. Pash, *Alsos Mission*, 86–87.

119. Pash to Chief, "Progress Report No. 7."

120. Pash, *Alsos Mission*, 99.

121. Pash, *Alsos Mission*, 87.

122. Goudsmit, *Alsos*, 66.

123. Pash, *Alsos Mission*, 130.

4. Transitions

1. Boris Pash, *The Alsos Mission* (New York: Award House, 1969), 140–42; Samuel A. Goudsmit, *Alsos*, vol. 1 of *The History of Modern Physics, 1800–1950* (New York: Henry Schuman, 1947), 66–67.

2. Robert Furman to Leslie Groves, November 29, 1944, Correspondence ("Top Secret") of the Manhattan Engineer District, 1942–1946, RG 77, M1109, roll 3, National Archives and Records Administration II, College Park, MD (hereafter cited as NARA II).

3. Pash does not provide this scientist's full name.

4. Pash, *Alsos Mission*, 151–54; Goudsmit, *Alsos*, 66–67.

5. Pash, *Alsos Mission*, 156–57; Goudsmit, *Alsos*, 69.

6. Pash, *Alsos Mission*, 157.

7. Goudsmit, *Alsos*, 69–71.

8. Goudsmit, *Alsos*, 70.

9. Goudsmit, *Alsos*, 108.

10. Pash, *Alsos Mission*, 157.

11. Pash, *Alsos Mission*, 159.

12. Vannevar Bush, *Pieces of the Action* (New York: William Morrow, 1970), 115.

13. Vannevar Bush, *Modern Arms and Free Men: A Discussion of the Role of Science in Preserving Democracy* (New York: Simon and Schuster, 1949), 206.

14. Bush, *Pieces of the Action*, 115.

15. Leslie Groves to Clayton Bissell, March 16, 1945, Correspondence ("Top Secret") of the Manhattan Engineer District, 1942–1946, RG 77, M1109, roll 4, NARA II.

16. Leslie R. Groves, *Now It Can Be Told: The Story of the Manhattan Project* (New York: Da Capo, 1975), 222.

17. Nicholas Dawidoff, *The Catcher Was a Spy* (New York: Vintage Books, 1994), 202–10.

18. Robert Furman to Leslie Groves, "Alsos Mission," September 5, 1944, Correspondence ("Top Secret") of the Manhattan Engineer District, 1942–1946, RG 77, M1109, roll 4, NARA II.

19. Samuel Goudsmit to his wife and daughter, December 10, 1944, Samuel A. Goudsmit Papers, American Institute of Physics Collections, Niels Bohr Library & Archives, http://www.aip.org/history/nbl/collections/goudsmit/index.html.

20. Pash, *Alsos Mission*, 164–65.

21. "Meeting of MIS—ALSOS Advisory Committee," December 16, 1944, Correspondence ("Top Secret") of the Manhattan Engineer District, 1942–1946, RG 77, M1109, roll 4, NARA II.

22. "Meeting of MIS—ALSOS Advisory Committee."

23. Boris Pash to Chief, Military Intelligence Service, "Current Organizational Disposition, Alsos Mission, European Theater of Operations," April 15, 1945, Correspondence ("Top Secret") of the Manhattan Engineer District, 1942–1946, RG 77, M1109, roll 4, NARA II.

24. Pash, *Alsos Mission*, 165.

25. Pash, *Alsos Mission*, 159.

26. There are many books available on the Soviet bomb program, some of them more recent, but the best two continue to be Richard Rhodes's *Dark Sun* and David Holloway's *Stalin and the Bomb*.

27. MED Counterintelligence, "Summary: Russian Situation," May 13, 1945, Records of the Office of the Commanding General, Manhattan Project, RG 77, box 13, NARA II.

28. Groves, *Now It Can Be Told*, 141.

29. John Lansdale Jr., "Military Service" (unpublished manuscript, 1987), Wood Library–Museum of Anesthesiology, 28–29, http://woodlibrarymuseum.org/ebooks/item/157/lansdale-john-jr-military-service.

30. Leslie Groves to Henry Wallace, Henry Stimson, and George Marshall, "Present Status and Future Program," August 23, 1943, Correspondence ("Top Secret") of the Manhattan Engineer District, 1942–1946, RG 77, M1109, roll 5, NARA II.

31. MED Counterintelligence, "Summary: Russian Situation."

32. J. Edgar Hoover to Harry Hopkins, February 9, 1945, Correspondence ("Top Secret") of the Manhattan Engineer District, 1942–1946, RG 77, M1109, roll 2, NARA II.

33. Groves to Wallace et al., "Present Status."

34. Richard Rhodes, *Dark Sun: The Making of the Hydrogen Bomb* (New York: Simon and Schuster, 1995), 100.

35. Groves to Wallace et al., "Present Status."

36. Lansdale, "Military Service," 7.

37. Leslie Groves to James Byrnes, May 15, 1945, Correspondence ("Top Secret") of the Manhattan Engineer District, 1942–1946, RG 77, M1109, roll 3, NARA II; "Transcription of Lansdale's Notes," November 29, 1944, Correspondence ("Top Secret") of the Manhattan Engineer District, 1942–1946, RG 77, M1109, roll 3, NARA II.

38. "Interview with Professor F. Joliot, London, September 5th and 7th, 1944," Office of the Chief of Engineers, S-1 History of Atomic Bomb, RG 77, entry A1 20, container 148, NARA II.

39. Leslie Groves to Henry Stimson, May 13, 1945, Harrison-Bundy Files Relating to the Development of the Atomic Bomb, 1942–1946, RG 77, M1108, roll 2, NARA II.

40. Central Intelligence Group, Intelligence Report, "Professor and Madame Joliot-Curie," January 7, 1947, CIA-RDP82-00457R000200470001-1, CIA Records Search Tool (CREST), NARA II.

41. Groves was correct in this assumption: Frédéric Joliot-Curie sent everything he could to his Soviet counterparts.

42. Leslie Groves to George Marshall, March 7, 1945, Correspondence ("Top Secret") of the Manhattan Engineer District, 1942–1946, RG 77, M1109, roll 1, NARA II.

43. Groves, *Now It Can Be Told*, 231.

44. George Marshall to Carl Spaatz, March 7, 1945, Correspondence ("Top Secret") of the Manhattan Engineer District, 1942–1946, RG 77, M1109, roll 1, NARA II.

45. The 1,684 tons of munitions consisted of 1,506 tons of high-explosive and 178 tons of incendiary bombs. Groves, *Now It Can Be Told*, 230–31.

46. Carl Spaatz to George Marshall, March 19, 1945, Correspondence ("Top Secret") of the Manhattan Engineer District, 1942–1946, RG 77, M1109, roll 1, NARA II.

47. Spaatz to Marshall, March 19, 1945.

48. Groves, *Now It Can Be Told*, 231.

49. Goudsmit, *Alsos*, 78.

50. Groves, *Now It Can Be Told*, 231.

51. Goudsmit, *Alsos*, 78.

52. Groves, *Now It Can Be Told*, 231.

53. Pash to Chief, "Current Organizational Disposition."

54. Goudsmit, *Alsos*, 89.

55. Goudsmit, *Alsos*, 87–90.

56. Groves, *Now It Can Be Told*, 234.

57. Leslie Groves to Henry Stimson, April 3, 1945, Correspondence ("Top Secret") of the Manhattan Engineer District, 1942–1946, RG 77, M1109, roll 1, NARA II.

58. Henry Stimson to Edward Stettinius, "Establishment of a French Zone of Occupation in Europe," April 3, 1945, Correspondence ("Top Secret") of the Manhattan Engineer District, 1942–1946, RG 77, M1109, roll 1, NARA II.

59. Groves (and several others, including FDR—see chapter 2) believed the quickest way to ensure that vital secret information would find its way to the enemy was to give it to the State Department.

60. John Lansdale, "Operation Harborage," draft report for Groves, dictated July 10, 1946, Correspondence ("Top Secret") of the Manhattan Engineer District, 1942–1946, RG 77, M1109, roll 1, NARA II.

61. Groves, *Now It Can Be Told*, 234.

62. Goudsmit, *Alsos*, 235.

63. Lansdale, "Operation Harborage."

64. Lansdale, "Operation Harborage."

65. John Lansdale to Leslie Groves, "ETO," May 5, 1945, Correspondence ("Top Secret") of the Manhattan Engineer District, 1942–1946, RG 77, M1109, roll 1, NARA II.

66. Lansdale to Groves, "ETO."

67. Lansdale to Groves, "ETO."

68. Leslie Groves to George Marshall, April 23, 1945, Correspondence ("Top Secret") of the Manhattan Engineer District, 1942–1946, RG 77, M1109, roll 2, NARA II.

69. Groves, *Now It Can Be Told*, 236–37.

70. John Lansdale, "Capture of Material," draft report, dictated July 10, 1946, Correspondence ("Top Secret") of the Manhattan Engineer District, 1942–1946, RG 77, M1109, roll 2, NARA II.

71. Lansdale, "Capture of Material."

72. Lansdale gave two different titles in two different documents.

73. Lansdale, "Capture of Material"; Lansdale, "Operation Harborage."

74. Lansdale, "Operation Harborage."

75. Groves insists it was the 1279th, while Pash's and Lansdale's recollections state that it was the 1269th Engineer Combat Battalion.

76. Pash, *Alsos Mission*, 206.

77. Lansdale, "Operation Harborage."

78. Lansdale, "Military Service," 62.

79. Lansdale to Groves, "ETO."

80. Lansdale, "Operation Harborage."

81. Lansdale, "Operation Harborage."

82. Groves, *Now It Can Be Told*, 242.

83. Richard Ham to Boris Pash, "Munich Operation," May 12, 1945, Correspondence ("Top Secret") of the Manhattan Engineer District, 1942–1946, RG 77, M1109, roll 4, NARA II.

84. Groves wrote that it was an eight-man group, Goudsmit a six-man group. The official report of the mission, written by Pash, notes that it was an eleven-man group.

85. Boris Pash to Chief, Military Intelligence Service, War Department, "Alpino Operation," May 18, 1945, Correspondence ("Top Secret") of the Manhattan Engineer District, 1942–1946, RG 77, M1109, roll 4, NARA II.

86. Pash to Chief, "Alpino Operation."

87. Pash to Chief, "Alpino Operation."

88. Groves, *Now It Can Be Told*, 243–44.

89. Pash, *Alsos Mission*, 242.

90. Groves, *Now It Can Be Told*, 248–49.

91. Charles Frank, ed., *Operation Epsilon: The Farm Hall Transcripts* (Berkeley: University of California Press, 1993), 203.

92. Frank, *Operation Epsilon*, 172.

93. Groves, *Now It Can Be Told*, 338.

94. Groves, *Now It Can Be Told*, 340.

5. Regression

1. Allen Dulles, *The Craft of Intelligence* (New York: Signet Books, 1965), 34.

2. The FBI was responsible for all foreign intelligence in the Western Hemisphere.

3. Harry S. Truman, *Memoirs of Harry S. Truman*, vol. 1, *Year of Decisions* (New York: Doubleday, 1955), 99.

4. "Executive Order 9621," September 20, 1945, U.S. Department of State, Office of the Historian, *Foreign Relations of the United States, 1945–1950, Emergence of the Intelligence Establishment* (hereafter cited as *FRUS, 1945–1950, Emergence*), document 14, https://history.state.gov/historicaldocuments/frus1945-50Intel/d14.

5. John McCloy to John Magruder, September 26, 1945, *FRUS, 1945–1950, Emergence*, document 95, https://history.state.gov/historicaldocuments/frus1945-50Intel/d95.

6. John Magruder to John McCloy, October 9, 1945, *FRUS, 1945–1950, Emergence*, document 96, https://history.state.gov/historicaldocuments/frus1945-50Intel/d96.

7. John Magruder to John McCloy, October 25, 1945, *FRUS, 1945–1950, Emergence*, document 97, https://history.state.gov/historicaldocuments/frus1945-50Intel/d97.

8. Adm. William Leahy became the president's representative.

9. Harry Truman to the Secretaries of State, War, and the Navy, January 22, 1946, CIA-RDP85S00362R000700130001-9, CIA Records Search Tool (CREST), National Archives and Records Administration II, College Park, MD (hereafter cited as NARA II).

10. Truman to the Secretaries of State, War, and the Navy, January 22, 1946.

11. Memorandum by the Director of Central Intelligence, Sidney Souers, CIG 8, "Development of Intelligence on U.S.S.R.," April 29, 1946, *FRUS, 1945–1950, Emergence*, document 148, https://history.state.gov/historicaldocuments/frus1945-50Intel/d148.

12. The act's long title is: "An Act to promote the national security by providing for a Secretary of Defense; for a National Military Establishment; for a Department of the Army, a Department of the Navy, and a Department of the Air Force; and for the coordination of the activities of the National Military Establishment with other departments and agencies of the Government concerned with the national security."

13. National Security Act of 1947, https://www.intelligence.senate.gov/sites/default/files/laws/nsact1947.pdf.

14. Leslie R. Groves, *Now It Can Be Told: The Story of the Manhattan Project* (New York: Da Capo, 1975), 376–77.

15. Vannevar Bush, *Science, the Endless Frontier: A Report to the President* (Washington, DC: U.S. Government Printing Office, 1945), 150–51.

16. Clarence G. Lasby, *Project Paperclip: German Scientists and the Cold War* (New York: Atheneum, 1971), 150.

17. Bush, *Science, the Endless Frontier*, 150–51.

18. Bush, *Science, the Endless Frontier*, 132.

19. Bush, *Science, the Endless Frontier*, 19.

20. Don K. Price, *Government and Science: Their Dynamic Relation in American Democracy* (New York: NYU Press, 1954), 76.

21. Bush, *Science, the Endless Frontier*, 13–14; Price, *Government and Science*, 32.

22. David Kaiser, "The Atomic Secret in Red Hands? American Suspicions of Theoretical Physicists during the Early Cold War," *Representations* 90, no. 1 (Spring 2005): 28.

23. Kaiser, "Atomic Secret," 29.

24. Kaiser, "Atomic Secret," 43.

25. "President Truman Speaks to the Scientists," *Bulletin of the Atomic Scientists* 4, no. 10 (October 1948): 291–93.

26. NIA Directive No. 7, "Coordination of Collection Activities," January 2, 1947, CIA-RDP85S00362R000700130001-9, CREST, NARA II.

27. NSC Intelligence Directive No. 2, "Coordination of Collection Activities Abroad," January 13, 1948, *FRUS, 1945–1950, Emergence*, document 425, https://history.state.gov/

historicaldocuments/frus1945-50Intel/d425; NSC Intelligence Directive No. 3, "Coordination of Intelligence Production," January 13, 1948, *FRUS, 1945–1950, Emergence*, document 426, https://history.state.gov/historicaldocuments/frus1945-50Intel/d426.

28. General Order No. 13, December 31, 1948, in Karl Weber, Historical Staff, Central Intelligence Agency, "The Office of Scientific Intelligence, 1949–68: Volume 1," June 1972, DD/S&T Historical Series, OSI-1.

29. "Office of Scientific Intelligence—Statement of Functions," February 7, 1949, in Weber, "Office of Scientific Intelligence," Annex A.

30. NSC Intelligence Directive No. 10, "Collection of Foreign Scientific and Technological Data," January 18, 1949, *FRUS, 1945–1950, Emergence*, document 429, https://history.state.gov/historicaldocuments/frus1945-50Intel/d429.

31. Research and Analysis Officer [name redacted] to William Donovan, "Influence of Atomic Bomb on Indirect Methods of Warfare," August 18, 1945, CIA-RDP84-00022R000200100026-4, CREST, NARA II.

32. Officer in Technical Section of OSS [name redacted] to William Donovan, "Influence of Atomic Bomb on Indirect Methods of Warfare," September 4, 1945, CIA-RDP84-00022R000300100007-4, CREST, NARA II.

33. Puleston was director of naval intelligence from June 1934 to April 1937.

34. William Puleston to Frederick Horne, "Post-War Organization of Naval Intelligence," September 22, 1945, CIA-RDP84-00022R000400070027-5, CREST, NARA II.

35. William H. Jackson to James Forrestal, November 14, 1945, CIA-RDP80R01731R002900440062-9, CREST, NARA II.

36. Clayton Bissell to George Marshall, "Scientific Military Intelligence Collection," August 25, 1945, Correspondence ("Top Secret") of the Manhattan Engineer District, 1942–1946, RG 77, M1109, roll 4, NARA II.

37. Vannevar Bush and James Conant to Henry Stimson, "Salient Points concerning Future International Handling of Subject of Atomic Bombs," September 30, 1944, Correspondence ("Top Secret") of the Manhattan Engineer District, 1942–1946, RG 77, M1109, roll 2, NARA II.

38. "Notes of the Interim Committee Meeting," May 31, 1945, Correspondence ("Top Secret") of the Manhattan Engineer District, 1942–1946, RG 77, M1109, roll 4, NARA II.

39. "Notes of the Interim Committee Meeting."

40. Vannevar Bush to James Byrnes, "Coming Conference with Mr. Atlee," November 5, 1945, Correspondence ("Top Secret") of the Manhattan Engineer District, 1942–1946, RG 77, M1109, roll 2, NARA II.

41. Jessica Wang, "Scientists and the Problem of the Public in Cold War America, 1945–1960," *Osiris* 17, no. 1 (2002): 328–29.

42. Harold Urey, "A Scientist Views the World Situation," *Bulletin of the Atomic Scientists* 1, no. 5 (February 15, 1946): 4.

43. "Senate Hearings on Atomic Energy," *Bulletin of the Atomic Scientists* 1, no. 2 (December 24, 1945): 3.

44. Albert Einstein, interview with Michael Amrine, "The Real Problem Is in the Hearts of Men," *New York Times Magazine*, June 23, 1946, SM4.

45. Glenn T. Seaborg, *Adventures in the Atomic Age: From Watts to Washington* (New York: Farrar, Straus and Giroux, 2001), 138–39.

46. Robert Gilpin, *American Scientists and Nuclear Weapons Policy* (Princeton, NJ: Princeton University Press, 1962), 53–54.

47. Gilpin, *American Scientists*, 60–61.

48. Leslie Groves to George Harrison, "Memorandum to Members of Interim Committee from V. Bush and J. B. Conant," July 25, 1945, Correspondence ("Top Secret") of the Manhattan Engineer District, 1942–1946, RG 77, M1109, roll 2, NARA II.

49. W. Averell Harriman to James Byrnes, November 19, 1945, Harrison-Bundy Files Relating to the Development of the Atomic Bomb, 1942–1946, RG 77, M1108, roll 2, NARA II.

50. Laurence Steinhardt [ambassador to Czechoslovakia] to James Byrnes, November 19, 1945, Harrison-Bundy Files Relating to the Development of the Atomic Bomb, 1942–1946, RG 77, M1108, roll 2, NARA II.

51. Roger Makins to Leslie Groves, November 7, 1945, Correspondence ("Top Secret") of the Manhattan Engineer District, 1942–1946, RG 77, M1109, roll 2, NARA II.

52. Henry Lowenhaupt, "On the Soviet Nuclear Scent," *Studies in Intelligence* 11, no. 4 (Fall 1967): 13–15. Lowenhaupt had obtained a doctorate in chemistry from Yale in 1943. During his time at Yale he had worked part time on uranium enrichment by chemical methods. After basic training at Oak Ridge, he was assigned to work for Groves in Washington, and in 1945 began to focus on foreign nuclear-related activities. In late 1946, Lowenhaupt, now a civilian, was still at the beginning of a long and distinguished career in intelligence.

53. Lowenhaupt, "On the Soviet Nuclear Scent," 15.

54. Weber, "Office of Scientific Intelligence," 6.

55. "Coordination of Intelligence Activities Related to Foreign Atomic Energy Developments and Potentialities," proposed NIA directive, August 13, 1946, CIA-RDP85S00362R000700100005-8, CREST, NARA II.

56. Lay would become executive secretary of the National Security Council from 1950 to 1961. He then was Allen Dulles's deputy assistant at the CIA, and finally the executive secretary of the U.S. Intelligence Board.

57. James Lay Jr. to Hoyt Vandenberg, "Coordination of Intelligence Activities Related to Foreign Atomic Energy Developments and Potentialities," August 21, 1946, CIA-RDP85S00362R000700100004-9, CREST, NARA II.

58. Minutes of the Sixth Meeting of the National Intelligence Authority, August 21, 1946, *FRUS, 1945–1950, Emergence*, document 163, https://history.state.gov/historicaldocuments/frus1945-50Intel/d163.

59. Minutes of the Sixth Meeting of the National Intelligence Authority.

60. Minutes of the Sixth Meeting of the National Intelligence Authority.

61. William Leahy to Harry Truman, telegram, August 21, 1946, *FRUS, 1945–1950, Emergence*, document 164, https://history.state.gov/historicaldocuments/frus1945-50Intel/d164.

62. Leahy to Truman, August 21, 1946, note 1.

63. Leslie Groves to the Atomic Energy Commission, November 21, 1946, *FRUS, 1945–1950, Emergence*, document 177, https://history.state.gov/historicaldocuments/frus1945-50Intel/d177.

64. Minutes of the Ninth Meeting of the National Intelligence Authority, February 12, 1947, *FRUS, 1945–1950, Emergence*, document 185, https://history.state.gov/historicaldocuments/frus1945-50Intel/d185.

65. NIA Directive No. 9, "Coordination of Intelligence Activities Related to Foreign Atomic Energy Developments and Potentialities," April 18, 1947, *FRUS, 1945–1950, Emergence*, document 194, https://history.state.gov/historicaldocuments/frus1945-50Intel/d194.

66. E. K. Wright, "Establishment and Functions of the Nuclear Energy Group, Scientific Branch, Office of Reports and Estimates," March 28, 1947, *FRUS, 1945–1950, Emergence*, document 191, https://history.state.gov/historicaldocuments/frus1945-50Intel/d191.

67. ORE 3/1, "Soviet Capabilities for the Development and Production of Certain Types of Weapons and Equipment," October 31, 1946, CIA-RDP67-00059A000200130011-3, CREST, NARA II.

68. ORE 3/1, "Soviet Capabilities for the Development and Production."

69. "Personnel of Scientific Intelligence Mission," August 22, 1944, and August 31, 1944, Correspondence ("Top Secret") of the Manhattan Engineer District, 1942–1946, RG 77, M1109, roll 4, NARA II.

70. Ronald Doel and Allan Needell, "Science, Scientists, and the CIA: Balancing International Ideals, National Needs, and Professional Opportunities," in *Eternal Vigilance? 50 Years of the CIA*, ed. Rhodri Jeffreys-Jones and Christopher Andrew (London: Frank Cass, 1997), 64–66.

71. Ralph Clark to Vannevar Bush, "CIA Situation," December 3, 1947, *FRUS, 1945–1950, Emergence*, document 333, enclosure 1, https://history.state.gov/historicaldocuments/frus1945-50Intel/d333.

72. Stephen Penrose, "Report on CIA," January 2, 1948, *FRUS, 1945–1950, Emergence*, document 338, enclosure 1, https://history.state.gov/historicaldocuments/frus1945-50Intel/d338.

73. Doel and Needell, "Science, Scientists, and the CIA," 64.

74. Chief of the Intelligence Section (Beckler) to Vannevar Bush, "The Critical Situation with Regard to Atomic Energy Intelligence," December 2, 1947, *FRUS, 1945–1950, Emergence*, document 333, enclosure 2, https://history.state.gov/historicaldocuments/frus1945-50Intel/d333.

75. Weber, "Office of Scientific Intelligence," 14.

76. Roscoe Hillenkoetter to Harry Truman, "Estimate of the Status of the Russian Atomic Energy Project," July 6, 1948, National Security Archive, https://nsarchive2.gwu.edu//nukevault/ebb286/doc03.PDF.

77. Doel and Needell, "Science, Scientists, and the CIA," 66–67.

78. "Status of the U.S.S.R. Atomic Energy Project," July 1, 1949, Records of Headquarters, U.S. Air Force (Air Staff), Deputy Chief of Staff for Operations, Directorate of Intelligence, July 1945–December 1954, RG 341, box 45, NARA II.

79. Intelligence Memorandum No. 225, "Estimate of Status of Atomic Warfare in the USSR," September 20, 1949, National Security Archive, https://nsarchive2.gwu.edu/radiation/dir/mstreet/commeet/meet6/brief6/tab_h/br6h1e.txt.

80. Willard Machle to Roscoe Hillenkoetter, "Inability of OSI to Accomplish Its Mission," September 29, 1949, *FRUS, 1945–1950, Emergence*, document 399, https://history.state.gov/historicaldocuments/frus1945-50Intel/d399.

81. Machle to Hillenkoetter, "Inability of OSI."

82. Machle to Hillenkoetter, "Inability of OSI."

6. Whistling in the Dark

1. Herbert York, *Race to Oblivion: A Participant's View of the Arms Race* (New York: Simon and Schuster, 1970), 108.

2. Richard Rhodes, *Dark Sun: The Making of the Hydrogen Bomb* (New York: Simon and Schuster, 1995), 368–73.

3. Joint Committee on Atomic Energy, "Report of the Central Intelligence Agency," October 17, 1949, Records of the Joint Committee on Atomic Energy, JCAE Transcripts, RG 128, box 3, National Archives and Records Administration II, College Park, MD (hereafter cited as NARA II).

4. John Lansdale Jr., "Military Service" (unpublished manuscript, 1987), Wood Library–Museum of Anesthesiology, http://woodlibrarymuseum.org/ebooks/item/157/lansdale-john-jr-military-service, 67.

5. "Notes of a Meeting on the Smyth Report in the Office of the Secretary of War," August 2, 1945, Correspondence ("Top Secret") of the Manhattan Engineer District, 1942–1946, RG 77, M1109, roll 2, NARA II.

6. "Notes of a Meeting on the Smyth Report."

7. Leslie Groves, "The Atom General Answers His Critics," *Saturday Evening Post*, June 19, 1948, 101.

8. "Bomb Was Predicted for 1952," *New York Times*, September 24, 1949. The *Times* was surprisingly close to reality with this argument. As chapter 4 shows, the Soviets had immediately understood the implications of fission, and while the German invasion interrupted their research, they began in earnest in 1943 under Igor Kurchatov, and then initiated a crash program after Trinity in 1945.

9. J. Edgar Hoover, "The Crime of the Century: The Case of the A-bomb Spies," *Reader's Digest*, May 1951, 158.

10. C. F. Trussell, "Ex-Major Says Hopkins Sped Uranium to Soviet in 1943; Wallace Named, Denies Role," *New York Times*, December 6, 1949.

11. Trussell, "Ex-Major."

12. Bernard Brodie, "What Is the Outlook Now?," *Bulletin of the Atomic Scientists* 5, no. 10 (October 1949): 268.

13. Eugene Rabinowitch, "Forewarned—But Not Forearmed," *Bulletin of the Atomic Scientists* 5, no. 10 (October 1949): 274.

14. Jack Raymond, "German Scientists Held Aiding Soviet," *New York Times*, September 24, 1949.

15. Clarence G. Lasby, *Project Paperclip: German Scientists and the Cold War* (New York: Atheneum, 1971), 6–7.

16. Franck had won the 1925 Nobel Prize in Physics for his confirmation of the Bohr model of the atom.

17. Sterns had been asked by Arthur Compton in 1942 to investigate the possibility of a German radiological attack (see chapter 2).

18. James Franck, Donald Hughes, J. J. Nickson, Eugene Rabinowitch, J. C. Sterns, Glenn Seaborg, and Leo Szilard., "Political and Social Problems," June 1945, from "Competitive Advantage of the United States over Other Nations with Respect to the Atomic Bomb: Extracts from the Records of the Interim Committee," Harrison-Bundy Files Relating to the Development of the Atomic Bomb, 1942–1946, RG 77, M1108, roll 2, NARA II.

19. Glenn T. Seaborg, *Adventures in the Atomic Age: From Watts to Washington* (New York: Farrar, Straus and Giroux, 2001), 141.

20. "Memorandum Prepared, and Subscribed To, by 300 Civilian Scientists at the New Mexico Laboratory, Dated 7 September 1945, as Transmitted by Dr. Oppenheimer to Mr. Harrison on 9 October 1945," from "Competitive Advantage of the United States over Other Nations with Respect to the Atomic Bomb: Extracts from the Records of the Interim Committee," Harrison-Bundy Files Relating to the Development of the Atomic Bomb, 1942–1946, RG 77, M1108, roll 2, NARA II.

21. "Did the Soviet Bomb Come Sooner Than Expected?," *Bulletin of the Atomic Scientists* 5, no. 10 (October 1949): 264.

22. Harrison Brown, *Must Destruction Be Our Destiny?* (New York: Simon and Schuster, 1946), 26.

23. "Russia and the Atomic Bomb," *Bulletin of the Atomic Scientists* 1, no. 5 (February 15, 1946), 10–11.

24. I compiled this information on Soviet scientists from a variety of sources, most notably Rhodes, *Dark Sun*; David Holloway, *Stalin and the Bomb: The Soviet Union and Atomic Energy, 1939–1956* (New Haven, CT: Yale University Press, 1996); "Russia and the Atomic Bomb"; Arnold Kramish, *Atomic Energy in the Soviet Union* (Stanford, CA: Stanford University Press, 1959); P. M. S. Blackett, *Atomic Weapons and East-West Relations* (Cambridge: Cambridge University Press, 1956); P. M. S. Blackett, *Studies of War: Nuclear and Conventional* (New York: Hill and Wang, 1962); Bernard Brodie, ed., *The Absolute Weapon: Atomic Power and World Order* (New York: Harcourt, 1946); and the oral histories of Luiz Alvarez, Hans Bethe, P. M. S. Blackett, Vannevar Bush, Robert Furman, Samuel Goudsmit, Philip Morrison, J. Robert Oppenheimer, Edward Teller, and Herbert York (AIP Physics Oral Histories, Niels Bohr Library & Archives).

25. Uranium undergoes extremely slow natural fission, without being exposed to any neutron bombardment.

26. Victor Weisskopf, *The Joy of Insight: Passions of a Physicist* (New York: Basic Books, 1991), 99.

27. York, *Race to Oblivion*, 107.

28. John Medaris and Arthur Gordon, *Countdown for Decision* (New York: G. P. Putnam's Sons, 1960), 53.

29. York, *Race to Oblivion*, 107; Rhodes, *Dark Sun*, 373.

30. "Did the Soviet Bomb Come Sooner Than Expected?," 262.

31. It might have been even earlier, but the first document available is from May 21, 1945.

32. Transcript of telephone conversation between Leslie Groves and G. M. Read, May 21, 1945, Correspondence ("Top Secret") of the Manhattan Engineer District, 1942–1946, RG 77, M1109, roll 2, NARA II.

33. "Notes of the Interim Committee Meeting, Friday, 1 June 1945," Correspondence ("Top Secret") of the Manhattan Engineer District, 1942–1946, RG 77, M1109, roll 4, NARA II.

34. "Notes of the Interim Committee Meeting, Friday, 1 June 1945."

35. Groves, "Atom General," 15–16, 100–102.

36. Hanson W. Baldwin, "Has Russia the Atomic Bomb?—Probably Not," *New York Times*, November 9, 1947.

37. Baldwin, "Has Russia the Atomic Bomb?"

38. Baldwin, "Has Russia the Atomic Bomb?"

39. Or "high-grade" uranium, defined as ore with a uranium content of 50 percent or greater.

40. Agreement of Trust signed by Franklin D. Roosevelt and Winston Churchill, June 13, 1944, Correspondence ("Top Secret") of the Manhattan Engineer District, 1942–1946, RG 77, M1109, roll 3, NARA II.

41. Robert Patterson to Leslie Groves, "Delegation of Authority under Executive Order #9001," Correspondence ("Top Secret") of the Manhattan Engineer District, 1942–1946, RG 77, M1109, roll 3, NARA II.

42. Donald P. Steury, "How the CIA Missed Stalin's Bomb: Dissecting Soviet Analysis, 1946–50," Central Intelligence Agency Center for the Study of Intelligence, last updated June 26, 2008, https://www.cia.gov/library/center-for-the-study-of-intelligence/csi-publications/csi-studies/studies/vol49no1/html_files/stalins_bomb_3.html.

43. Henry Lowenhaupt, "Chasing Bitterfeld Calcium," *Studies in Intelligence* 17, no. 1 (Spring 1973), CIA-RDP78T03194A000400010002-9, CIA Records Search Tool (CREST), NARA II.

44. For further information about the relationship between the Soviet system and science, see Lawrence Badash, *Kapitza, Rutherford, and the Kremlin* (New Haven, CT: Yale University Press, 1985); Vadim Birstein, *The Perversion of Knowledge: The True Story of Soviet Science* (New York: Basic Books, 2004); Thomas Cochran, Robert Norris, and Oleg Bukharin, *Making the Russian Bomb: From Stalin to Yeltsin* (Boulder, CO: Westview, 1995); Stephen Fortescue, *Science Policy in the Soviet Union* (London: Routledge, 1990); Loren Graham, *Science in Russia and the Soviet Union* (Cambridge: Cambridge University Press, 1994); Paul Josephson, *Physics and Politics in Revolutionary Russia* (Berkeley: University of California Press, 1991); Paul Josephson, *Red Atom: Russia's Nuclear Power Program from Stalin to Today* (New York: W. H. Freeman, 2000); A. B. Kojevnikov, *Stalin's Great Science: The Times and Adventures of Soviet Physicists* (London: Imperial College Press, 2004); Kramish, *Atomic Energy*; Bruce Parrott, *Politics and Technology in the Soviet Union* (Cambridge, MA: MIT Press, 1983); Albert Parry, *Peter Kapitza on Life and Science* (New York: Macmillan, 1968); Albert Parry, *The Russian Scientist* (New York: Macmillan, 1973); Ethan Pollock, *Stalin and the Soviet Science Wars* (Princeton, NJ: Princeton University Press, 2006); and Valery Soyfer, *Lysenko and the Tragedy of Soviet Science* (New Brunswick, NJ: Rutgers University Press, 1994).

45. Waldemar Kaempffert, "Science—and Ideology—in Soviet Russia," *New York Times*, September 15, 1946.

46. Karl Sax, "Soviet Science and Political Philosophy," *Scientific Monthly* 65, no. 1 (July 1947): 43–47.

47. Sax, "Soviet Science."

48. Vladimir Asmous, "Freedom of Science in Soviet Union," *Science*, n.s., 103, no. 2670 (March 1, 1946): 281–82.

49. H. J. Muller, "The Crushing of Genetics in the USSR," *Bulletin of the Atomic Scientists* 4, no. 12 (December 1948): 369–71.

50. "The Purge of Genetics in the Soviet Union," *Bulletin of the Atomic Scientists* 5, no. 5 (May 1949): 130.

51. Lewis Feuer, *Karl Marx and Friedrich Engels: Basic Writings on Politics and Philosophy* (New York: Doubleday, 1959).

52. Lewis Feuer, "Dialectical Materialism and Soviet Science," *Philosophy of Science* 16, no. 2 (April 1949): 116.

53. "Morgan" was Thomas Hunt Morgan, an American geneticist who won the Nobel Prize in Medicine in 1933 for his discovery of how chromosomes work in heredity.

54. Feuer, "Dialectical Materialism," 117.

55. "Did the Soviet Bomb Come Sooner Than Expected?," 264.

56. Samuel A. Goudsmit, *Alsos*, vol. 1 of *The History of Modern Physics, 1800–1950* (New York: Henry Schuman, 1947), 235.

57. Goudsmit, *Alsos*, xxvii–xxviii.

58. "Notes of the Interim Committee Meeting, Thursday, 31 May 1945," Correspondence ("Top Secret") of the Manhattan Engineer District, 1942–1946, RG 77, M1109, roll 4, NARA II.

59. Vannevar Bush, *Modern Arms and Free Men: A Discussion of the Role of Science in Preserving Democracy* (New York: Simon and Schuster, 1949), 93.

60. Bush, *Modern Arms*, 206.

61. Bush, *Modern Arms*, 209.

62. Bush, *Modern Arms*, 210.

Conclusion

1. "OSI Survey Report," February 1, 1952, CIA Freedom of Information Act Electronic Reading Room, Historical Collections, http://www.foia.cia.gov/sites/default/files/document_conversions/49/osi_survey_report.pdf.

2. For an in-depth analysis of early Cold War military policy, see David Alan Rosenberg, "The Origins of Overkill: Nuclear Weapons and American Strategy, 1945–1960," *International Security* 7, no. 4 (1983): 3–71.

Selected Bibliography

One of the most difficult aspects of any study of intelligence, particularly one that also deals with atomic weapons, is the issue of classification and secrecy. While seventy years have passed since the events in this book occurred, there are still documents that are unavailable to researchers due to their perceived importance to national security. Compounding this problem is the fact that many of the decisions of the U.S. atomic intelligence organization were not recorded. Leslie Groves was especially careful when it came to security and very often would transmit orders via word of mouth alone. This was the case for many of the highly secret missions of the MED intelligence team. In fact, there is no official document authorizing Groves to create his own intelligence organization. Army chief of staff George Marshall gave him verbal orders, in secret, and this oral agreement set in motion the creation of a formalized intelligence apparatus. What this means, in practice, is that this book depends heavily on the memoirs and oral interviews of key personnel. While this is not a perfect solution by any means, memoirs and oral interviews can mitigate classification issues and can provide context to events that cannot be fully understood through the documentary record alone. This book attempts to compensate for some of the problematic aspects of memoirs by using other sources to corroborate the information obtained from memoirs whenever possible.

Primary Sources

American Institute of Physics Collections, Niels Bohr Library & Archives, College Park, MD

Samuel A. Goudsmit Papers
Archive for the History of Quantum Physics

American Institute of Physics Oral Histories, Niels Bohr Library & Archives, College Park, MD

Luis Alvarez
Edoardo Amaldi
Robert Bacher
Hans Bethe
P. M. S. Blackett
Niels Bohr
Margrethe Norlund Bohr
Max Born
Norris Bradbury
Vannevar Bush
James Chadwick
John Cockcroft
Edward Condon
Peter Debye
Lee DuBridge
Richard Feynman
James Fisk
James Franck
Otto Frisch
Robert Furman
Wolfgang Gentner
Samuel Goudsmit
Paul Harteck
Werner Heisenberg
Gustav Hertz
Ernst Pascual Jordan
Lew Kowarski
Lise Meitner
Philip Morrison
Frank Oppenheimer
J. Robert Oppenheimer
Linus Pauling
Rudolf Peierls
Francis Perrin
Isidor Isaac Rabi
Nikolaus Riehl
Emilio Segré
Robert Serber
Edward Teller
Harold Urey
Victor Weisskopf
Carl Friedrich von Weizsäcker
Eugene Wigner
Herbert York

Library of Congress, Manuscript Division

Henry H. Arnold Papers
Vannevar Bush Papers
W. Averell Harriman Papers
Curtis E. LeMay Papers
Brien McMahon Papers
J. Robert Oppenheimer Papers
Isidor Isaac Rabi Papers
Glenn Theodore Seaborg Papers
Carl A. Spaatz Papers
Hoyt S. Vandenberg Papers

National Archives and Records Administration II, College Park, MD

Records of the Atomic Energy Commission (RG 326)
Bush-Conant Files Related to the Development of the Atomic Bomb (RG 227)

Records of the Office of Scientific Research and Development (RG 227)
Records of the War Department General and Special Staffs, Office of the Director of Intelligence, G-2 (RG 165)
Records of the Central Intelligence Agency (263)
Records of the Office of the Chief of Engineers (RG 77)
Records of the Manhattan Project (RG 77.11)
—Includes Correspondence ("Top Secret") of the Manhattan Engineer District
Leslie Groves Papers (RG 200)
Records of the National Security Council (RG 273)
General Records of the Department of State (RG 59)
Records of the U.S. Joint Chiefs of Staff (RG 218)
Records of the Office of Strategic Services (RG 226)
CIA Records Search Tool (CREST)

Online Resources

Niels Bohr Archive, https://www.nbarchive.dk
National Security Archive, https://nsarchive.gwu.edu
Central Intelligence Agency, Library, Freedom of Information Act Electronic Reading Room, https://www.cia.gov/library/readingroom
U.S. Department of State, Office of the Historian, https://history.state.gov
Federation of Atomic Scientists, Publications and Reports, https://www.fas.org/pubs/index.html
Alsos Digital Library for Nuclear Issues, http://alsos.wlu.edu
National Security Agency, Freedom of Information Act Electronic Reading Room, https://www.nsa.gov/resources/everyone/foia/reading-room
American Institute of Physics, Niels Bohr Library & Archives, https://www.aip.org/history-programs/niels-bohr-library

Journals/Periodicals

Atomic Scientists News
Bulletin of the American Physical Society
Bulletin of the Atomic Scientists
Christian Science Monitor
Diplomatic History
Foreign Affairs
Harper's
Intelligence and National Security
Life
The Nation
Nature
New York Times
Representations
Saturday Evening Post
Science

Science Newsletter
Time
Washington Post

Institutional Histories

American Institute of Physics Institutional Histories, Niels Bohr Library & Archives, College Park, MD

Argonne National Laboratory. "Twenty Years of Nuclear Progress," 1962. Call number IH13.

Bankoff, S. George. "S. George Bankoff Recollections of the First Hanford Nuclear Reactor, 1943–1945," 1995. Call number IH4074.

Columbia University. "The Columbia Physics Department: A Brief History," n.d. Call number IH4083.

Los Alamos National Laboratory. "Los Alamos, the Beginning of an Era, 1943–1945," 1984. Call number IH123.

Parker, Vincent Eveland. "The History, Mission, Organization and Present Program of ORINS [Oak Ridge Institute of Nuclear Studies]," 1964. Call number IH163.

University of California, Berkeley, Department of Physics. "Breaking Through: A Century of Physics at Berkeley," 2009. Call number IH2009-917.

Published Institutional Histories

Benson, Robert. *The Venona Story*. Fort Meade, MD: National Security Agency, Center for Cryptologic History, 2001.

Noyes, William, ed. *Chemistry: A History of the Chemistry Component of the NDRC, 1940–1946*. Boston: Atlantic–Little, Brown, 1948.

Rearden, Steven. *History of the Office of the Secretary of Defense*. Vol. 1, *The Formative Years, 1947–1950*. Washington, DC: Historical Office, Office of the Secretary of Defense, 1984.

Published Primary Documents

Badash, Lawrence, Joseph Hirschfelder, and Herbert Broida, eds. *Reminiscences of Los Alamos, 1943–1945*. Dordrecht, Netherlands: D. Reidel, 1980.

Benson, Robert Louis, and Michael Warner, eds. *Venona: Soviet Espionage and the American Response, 1939–1957*. Washington, DC: National Security Agency, Central Intelligence Agency, 1996.

Bernstein, Jeremy. *Hitler's Uranium Club: The Secret Recordings at Farm Hall.* Woodbury, NY: American Institute of Physics, 1996.

Cantelon, Philip, Richard Hewlett, and Robert Williams, eds. *The American Atom: A Documentary History of Nuclear Policies from the Discovery of Fission to the Present.* Philadelphia: University of Pennsylvania Press, 1991.

David, L. R., and I. A. Warheit. *German Reports on Atomic Energy: Bibliography of ALSOS Technical Reports (YID-3030).* Oak Ridge, TN: U.S. Atomic Energy Commission, 1952.

Dörries, Matthias, ed. *Michael Frayn's "Copenhagen" in Debate: Historical Essays and Documents on the 1941 Meeting between Niels Bohr and Werner Heisenberg.* Berkeley: Office for History of Science and Technology, University of California, 2005.

Ermenc, Joseph J., ed. *Atomic Bomb Scientists: Memoirs, 1939–1945.* Westport, CT: Meckler, 1989.

Merrill, Dennis, ed. *Documentary History of the Truman Presidency.* Vol. 21, *The Development of an Atomic Weapons Program following World War II.* Bethesda, MD: University Publications of America, 1998.

Rhodes, Richard, ed. *The Los Alamos Primer: The First Lectures on How to Build an Atomic Bomb.* Annotated by Robert Serber. Berkeley: University of California Press, 1992.

Smyth, Henry DeWolf. *Atomic Energy for Military Purposes: The Official Report on the Development of the Atomic Bomb under the Auspices of the United States Government, 1940–45.* Princeton, NJ: Princeton University Press, 1945.

Memoirs and Contemporary Sources

Acheson, Dean. *Present at the Creation: My Years in the State Department.* New York: W. W. Norton, 1969.

Allier, Jacques. "The First Atomic Piles and the French Effort." *Atomic Scientists News,* no. 11 (1953): 225–48.

Barnard, Chester, J. R. Oppenheimer, Charles A. Thomas, Harry A. Winne, and David E. Lilienthal. *A Report on the International Control of Atomic Energy.* Washington, DC: U.S. Department of State, 1946.

Baxter, James Phinney. *Scientists against Time.* Boston: Little, Brown, 1946.

Bethe, Hans. "Brighter Than a Thousand Suns." *Bulletin of the Atomic Scientists* 14, no. 10 (December 1958): 426–28.

Blackett, P. M. S. *Atomic Weapons and East-West Relations.* Cambridge: Cambridge University Press, 1956.

——. *Studies of War: Nuclear and Conventional.* New York: Hill and Wang, 1962.

Bohr, Niels. "Disintegration of Heavy Nuclei." *Nature* 143, no. 3617 (February 25, 1939): 330.

Born, Max. *The Born-Einstein Letters.* New York: Walker, 1971.

——. *My Life and Views.* New York: Scribner's, 1968.

——. *My Life: Recollections of a Nobel Laureate.* New York: Scribner's, 1978.

——. *Physics and Politics.* New York: Basic Books, 1962.

Bradley, Omar. *A Soldier's Story*. New York: Henry Holt, 1951.

Brodie, Bernard, ed. *The Absolute Weapon: Atomic Power and World Order*. New York: Harcourt, 1946.

Bundy, McGeorge. *Danger and Survival: Choices about the Bomb in the First Fifty Years*. New York: Random House, 1988.

Bush, Vannevar. *Modern Arms and Free Men: A Discussion of the Role of Science in Preserving Democracy*. New York: Simon and Schuster, 1949.

——. *Pieces of the Action*. New York: William Morrow, 1970.

——. *Science, the Endless Frontier: A Report to the President*. Washington, DC: U.S. Government Printing Office, 1945.

——. *Science Is Not Enough*. New York: William Morrow, 1965.

Compton, Arthur Holly. *Atomic Quest: A Personal Narrative*. New York: Oxford University Press, 1956.

Compton, Karl T. "If the Atomic Bomb Had Not Been Used." *Atlantic*, December 1946, 54–56.

——. "Organization of American Scientists for War, I." *Science* 98, no. 2535 (July 23, 1943): 71–76.

——. "Organization of American Scientists for War, II." *Science* 98, no. 2535 (July 23, 1943): 93–98.

Conant, James B. *My Several Lives: Memoirs of a Social Inventor*. New York: Harper and Row, 1970.

Copeland, G. H. "Nazi Science Secrets." *New York Times Magazine*, February 23, 1947, 33–34.

Dulles, Allen. *The Craft of Intelligence*. Guilford, CT: Lyons, 1963

——. *Germany's Underground*. New York: Macmillan, 1947.

Einstein, Albert, and Leopold Infeld. *The Evolution of Physics: From Early Concepts to Relativity and Quanta*. New York: Simon and Schuster, 1966. First published 1938 by Cambridge University Press.

Eliot, George Fielding. "Science and Foreign Policy." *Foreign Affairs* 23, no. 3 (April 1945): 378–87.

Fermi, Laura. *Atoms in the Family: My Life with Enrico Fermi*. Chicago: University of Chicago Press, 1954.

——. *Illustrious Immigrants: The Intellectual Migration from Europe, 1930–1941*. Chicago: University of Chicago Press, 1968.

Frisch, O. R. *What Little I Remember*. Cambridge: Cambridge University Press, 1979.

Frisch, O. R., F. A. Paneth, F. Laves, and P. Rosbaud, eds. *Trends in Atomic Physics*. New York: Inter-science, 1959.

Gamow, George. *Thirty Years That Shook Physics: The Story of Quantum Theory*. New York: Doubleday, 1966.

Goudsmit, Samuel A. *Alsos*. Vol. 1 of *The History of Modern Physics, 1800–1950*. New York: Henry Schuman, 1947.

——. "Nazis' Atomic Secrets." *Life*, October, 20, 1946, 123–34.

——. "Why the Nazis Did Not Get the Atomic Bomb." *Bulletin of the American Physical Society*, no. 22 (May 1947): 4, 22.

Groves, Leslie R. *Now It Can Be Told: The Story of the Manhattan Project*. New York: Da Capo, 1975.

Hahn, Otto. *A Scientific Autobiography*. New York: Scribner's, 1966.

——. *My Life*. New York: Herder and Herder, 1970.

Halban, Hans von, F. Joliot-Curie, and L. Kowarski. "Liberation of Neutrons in the Nuclear Explosion of Uranium." *Nature* 143, no. 3620 (March 18, 1939): 470–71.

——. "Number of Neutrons Liberated in the Nuclear Fission of Uranium." *Nature* 143, no. 3625 (April 22, 1939): 680–81.

Haukelid, Knut. *Skis against the Atom*. London: William Kimber, 1954.

Heisenberg, Elisabeth. *Inner Exile: Recollections of a Life with Werner Heisenberg*. Boston: Birkhauser, 1984.

Heisenberg, Werner. *Physics and Beyond: Encounters and Conversations*. New York: Harper and Row, 1971.

——. "Research in Germany on the Technical Application of Atomic Energy." *Nature* 160, no. 4059 (August 16, 1947): 211–15.

——. "The Third Reich and the Atomic Bomb." *Bulletin of the Atomic Scientists* 24, no. 6 (June 1968): 34–35.

Joint Committee on Atomic Energy, U.S. Congress. *Soviet Atomic Espionage*. Washington, DC: U.S. Government Printing Office, 1951.

Jones, R. V. *The Wizard War: British Scientific Intelligence, 1939–1945*. New York: Coward, McCann, and Geoghegan, 1978.

Kast, Fremont, and James Rosenzweig, eds. *Science, Technology, and Management*. New York: McGraw-Hill, 1963.

Laurence, William. *Dawn over Zero: The Story of the Atomic Bomb*. New York: Alfred A. Knopf, 1946.

——. *Men and Atoms: The Discovery, the Uses, and the Future of Atomic Energy*. New York: Simon and Schuster, 1959.

Lilienthal, David. *The Atomic Energy Year, 1945–1950: The Journals of David E. Lilienthal, Volume 2*. New York: Harper, 1964.

Masters, Dexter, and Katherine Way, eds. *One World or None*. New York: Whittlesey House, 1946.

Meitner, Lise. "Looking Back." *Bulletin of the Atomic Scientists* 20, no. 9 (November 1964): 2–7.

Meitner, Lise, and O. R. Frisch. "Disintegration of Uranium by Neutrons." *Nature* 143, no. 3615 (February 11, 1939): 330.

Morrison, Philip. "ALSOS: The Story of German Science." *Bulletin of the Atomic Scientists* 3, no. 12 (December 1947): 365.

Pash, Boris. *The Alsos Mission*. New York: Award House, 1969.

Peierls, Sir Rudolf. *Atomic Histories*. Woodbury, NY: American Institute of Physics, 1997.

——. *Bird of Passage: Recollections of a Physicist*. Princeton, NJ: Princeton University Press, 1985.

Roosevelt, Kermit. *War Report of the O.S.S.* New York: Walker, 1976.

Seaborg, Glenn T. *Adventures in the Atomic Age: From Watts to Washington*. New York: Farrar, Straus and Giroux, 2001.

Smith, Walter Bedell. *Eisenhower's Six Great Decisions: Europe, 1944–1945*. New York: Longman's Green, 1956.

Soddy, Frederick. *Radio-Activity: An Elementary Treatise*. London: "The Electrician," 1904.

Speer, Albert. *Inside the Third Reich*. New York: MacMillan, 1970.
Stimson, Henry. "The Bomb and the Opportunity." *Harper's*, March 1946, 204.
——. "The Decision to Use the Atomic Bomb." *Harper's*, February 1947, 97–107.
Stimson, Henry L., and McGeorge Bundy. *On Active Service in Peace and War*. New York: Harper, 1948.
Strauss, Lewis. *Men and Decisions*. New York: Doubleday, 1962.
Strong, Kenneth. *Intelligence at the Top: The Recollections of an Intelligence Officer*. New York: Doubleday, 1969.
Szilard, Leo. *The Collected Works: Scientific Papers*. Cambridge, MA: MIT Press, 1972.
——. "We Turned the Switch." *Nation*, December 22, 1945, 718–19.
U.S. Department of State. *The International Control of Atomic Energy: Growth of a Policy*. Washington, DC: U.S. Government Printing Office, 1946.
Weisskopf, Victor. *The Joy of Insight: Passions of a Physicist*. New York: Basic Books, 1991.
Winnacker, Karl, and Karl Wirtz. *Nuclear Energy in Germany*. La Grange Park, IL: American Nuclear Society, 1979.
York, Herbert. *The Advisors: Oppenheimer, Teller, and the Superbomb*. Palo Alto, CA: Stanford University Press, 1989.
——. *Arms and the Physicist*. Woodbury, NY: AIP Press, 1995.
——. *Race to Oblivion: A Participant's View of the Arms Race*. New York: Simon and Schuster, 1970.

Secondary Sources

Aczel, Amir D. *Uranium Wars: The Scientific Rivalry That Created the Nuclear Age*. New York: Palgrave Macmillan, 2009.
Allen, James. *Atomic Imperialism: The State, Monopoly, and the Bomb*. New York: International, 1952.
Alperovitz, Gar. *Atomic Diplomacy*. New York: Penguin, 1985.
——. *The Decision to Use the Atomic Bomb*. London: HarperCollins, 1995.
Alsop, Stewart, and Thomas Braden. *Sub Rosa: The OSS and American Espionage*. New York: Reynal and Hitchcock, 1946.
Bacher, Robert. *Robert Oppenheimer, 1904–1907*. Los Alamos, NM: Los Alamos Historical Society, 1999.
Badash, Lawrence. *Kapitza, Rutherford, and the Kremlin*. New Haven, CT: Yale University Press, 1985.
Bar-Zohar, Michael. *The Hunt for German Scientists*. New York: Hawthorn Books, 1967.
Bascomb, Neil. *The Winter Fortress: The Epic Mission to Sabotage Hitler's Atomic Bomb*. Boston: Houghton Mifflin Harcourt, 2016.
Bechhoefer, Bernard. *Postwar Negotiations for Arms Control*. Washington, DC: Brookings Institution, 1961.
Bergier, Jacques. *Secret Weapons—Secret Agents*. London: Hurst and Blackett, 1956.

Bernstein, Jeremy. *Hans Bethe: Prophet of Energy*. New York: Basic Books, 1980.
Beyerchen, Alan D. *Scientists under Hitler: Politics and the Physics Community in the Third Reich*. New Haven, CT: Yale University Press, 1977.
Biquard, Pierre. *Frédéric Joliot-Curie*. Greenwich, CT: Fawcett, 1966.
Bird, Kai, and Martin J. Sherwin. *American Prometheus: The Triumph and Tragedy of J. Robert Oppenheimer*. New York: Knopf, 2005.
Birstein, Vadim. *The Perversion of Knowledge: The True Story of Soviet Science*. New York: Basic Books, 2004.
Boyer, Paul. *By the Bomb's Early Light: American Thought and Culture at the Dawn of the Atomic Age*. Chapel Hill: University of North Carolina Press, 1985.
Bukharin, Oleg. "US Atomic Intelligence against the Soviet Target, 1945–1970." *Intelligence and National Security* 19, no. 4 (2004): 655–79.
Carson, Cathryn, and David Hollinger, eds. *Reappraising Oppenheimer: Centennial Studies and Reflections*. Berkeley: Office for History of Science and Technology, University of California, 2005.
Cassidy, David C. *Beyond Uncertainty: Heisenberg, Quantum Physics, and the Bomb*. New York: Bellevue Literary Press, 2009.
Cave Brown, Anthony. *The Last Hero: Wild Bill Donovan*. New York: Times Books, 1982.
Childs, Herbert. *An American Genius: The Life of Ernest Orlando Lawrence*. New York: Dutton, 1968.
Clark, Ronald W. *The Birth of the Bomb*. London: Scientific Book Club, 1961.
——. *The Greatest Power on Earth*. New York: Harper and Row, 1980.
——. *Tizard*. Cambridge, MA: MIT Press, 1965.
Cochran, Thomas, Robert Norris, and Oleg Bukharin. *Making the Russian Bomb: From Stalin to Yeltsin*. Boulder, CO: Westview, 1995.
Conant, Jennet. *109 East Palace: Robert Oppenheimer and the Secret City of Los Alamos*. New York: Simon and Schuster, 2005.
Corson, William R. *The Armies of Ignorance: The Rise of America's Intelligence Empire*. New York: Dial, 1977.
Craig, Campbell, and Sergey Radchenko. *The Atomic Bomb and the Origins of the Cold War*. New Haven, CT: Yale University Press, 2008.
Cropper, William H. *Great Physicists: The Life and Times of Leading Physicists from Galileo to Hawking*. New York: Oxford University Press, 2001.
Davis, Nuel Pharr. *Lawrence & Oppenheimer*. New York: Simon and Schuster, 1968.
Dawidoff, Nicholas. *The Catcher Was a Spy: The Mysterious Life of Moe Berg*. New York: Vintage Books, 1994.
Deacon, Richard. *A History of the British Secret Service*. London: Panther Books, 1980.
de Silva, Peer. *Sub Rosa: The CIA and the Uses of Intelligence*. New York: Times Books, 1978.
Dunlop, Richard. *Donovan: America's Master Spy*. Chicago: Rand McNally, 1982.
Dupree, A. Hunter. *Science in the Federal Government*. New York: Harper and Row, 1957.
Esterer, Arnulf, and Louise Esterer. *Prophet of the Atomic Age: Leo Szilard*. New York: Julian Messner, 1972.

Ford, Corey. *Donovan of OSS*. Boston: Little, Brown, 1970.
Ford, Corey, and Alastair MacBain. *Cloak and Dagger: The Secret Story of the O.S.S.* New York: Random House, 1946.
Fortescue, Stephen. *Science Policy in the Soviet Union*. London: Routledge, 1990.
Frayn, Michael. *Copenhagen*. New York: Anchor Books, 1998.
Frayn, Michael, and David Burke. *The Copenhagen Papers: An Intrigue*. New York: Metropolitan Books, 2000.
Freedman, Lawrence. *The Evolution of Nuclear Strategy*. New York: St. Martin's, 1983.
——. *U.S. Intelligence and the Soviet Strategic Threat*. Boulder, CO: Westview, 1977.
Gaddis, John Lewis. "Intelligence, Espionage, and Cold War Origins." *Diplomatic History* 13, no. 2 (1989): 191–212.
Gallagher, Thomas. *Assault in Norway: Sabotaging the Nazi Nuclear Bomb*. New York: Bantam Books, 1981.
Gertcher, Frank L. *Beyond Deterrence: The Political Economy of Nuclear Weapons*. Boulder, CO: Westview, 1990.
Gilpin, Robert. *American Scientists and Nuclear Weapons Policy*. Princeton, NJ: Princeton University Press, 1962.
Gimbel, John. *Science, Technology, and Reparations: Exploitation and Plunder in Postwar Germany*. Palo Alto, CA: Stanford University Press, 1990.
Gjelsvik, Tore. *Norwegian Resistance*. London: Hurst, 1979.
Goldschmidt, Bertrand. *Atomic Rivals: A Candid Memoir of Rivalries among the Allies over the Bomb*. New Brunswick, NJ: Rutgers University Press, 1990.
Goodman, Michael. *Spying on the Nuclear Bear: Anglo-American Intelligence and the Soviet Bomb*. Palo Alto, CA: Stanford University Press, 2007.
Gordin, Michael. *Five Days in August: How World War II Became a Nuclear War*. Princeton, NJ: Princeton University Press, 2007.
——. *Red Cloud at Dawn: Truman, Stalin, and the End of the Atomic Monopoly*. New York: Farrar, Straus and Giroux, 2009.
Gowing, Margaret. *Great Britain and Atomic Energy, 1939–1945*. London: Macmillan, 1964.
Graham, Loren. *Science in Russia and the Soviet Union*. Cambridge: Cambridge University Press, 1994.
Greene, Benjamin. *Eisenhower, Science Advice, and the Nuclear Test Ban Debate, 1945–1963*. Palo Alto, CA: Stanford University Press, 2007.
Groueff, Stephanie. *The Manhattan Project: The Untold Story of the Making of the Atomic Bomb*. New York: Bantam Books, 1967.
Haber, L. F. *The Poisonous Cloud*. Oxford: Oxford University Press, 1986.
Haines, Gerald, and Robert Leggett, eds. *Watching the Bear: Essays on CIA's Analysis of the Soviet Union*. Washington, DC: Center for the Study of Intelligence, Central Intelligence Agency, 2001.
Hart, B. H. Liddell. *History of the Second World War*. London: Cassells, 1970.
Hartcup, Guy. *The Effect of Science on the Second World War*. New York: St. Martin's, 2000.
Hartcup, Guy, and T. E. Allibone. *Cockcroft and the Atom*. Bristol: Adam Hilger, 1984.

Hasegawa, Tsuyoshi. *Racing the Enemy: Stalin, Truman, and the Surrender of Japan.* Cambridge, MA: Harvard University Press, 2005.

Hawkins, David, Edith Truslow, and Ralph Carlisle Smith. *Project Y: The Los Alamos Story.* Part 1, *Toward Trinity.* Los Angeles: Tomash, 1983.

Haynes, John Earl, and Harvey Klehr. *Venona: Decoding Soviet Espionage in America.* New Haven, CT: Yale University Press, 1999.

Helmreich, Jonathan. *Gathering Rare Ores: The Diplomacy of Uranium Acquisition, 1943–54.* Princeton, NJ: Princeton University Press, 1986.

Herken, Gregg. *Brotherhood of the Bomb: The Tangled Lives and Loyalties of Robert Oppenheimer, Ernest Lawrence, and Edward Teller.* New York: Henry Holt, 2002.

——. *The Winning Weapon: The Atomic Bomb and the Cold War, 1945–1950.* Princeton, NJ: Princeton University Press, 1988.

Hermann, Armin. *Werner Heisenberg, 1901–1976.* Bonn: Inter Nationes, 1976.

Hershberg, James. *James B. Conant: Harvard to Hiroshima and the Making of the Nuclear Age.* New York: Knopf, 1993.

Hewlett, Richard G., and Oscar E. Anderson Jr. *A History of the United States Atomic Energy Commission.* Vol. 1, *The New World, 1939–1946.* Berkeley: University of California Press, 1990.

Hewlett, Richard G., and Francis Duncan. *A History of the United States Atomic Energy Commission.* Vol. 2, *Atomic Shield, 1947–1952.* Berkeley: University of California Press, 1990.

Hindley, Meredith. *Destination Casablanca: Exile, Espionage, and the Battle for North Africa in World War II.* New York: Hachette, 2017.

Hinsley, F. H. *British Intelligence in the Second World War: Its Influence on Strategy and Operations.* 5 vols. New York: Cambridge University Press, 1979.

Hoddeson, Lillian, Paul Henriksen, Roger Meade, and Catherine Westfall. *Critical Assembly: A Technical History of Los Alamos during the Oppenheimer Years, 1943–1945.* New York: Cambridge University Press, 1993.

Holloway, David. *The Soviet Union and the Arms Race.* New Haven, CT: Yale University Press, 1983.

——. *Stalin and the Bomb: The Soviet Union and Atomic Energy, 1939–1956.* New Haven, CT: Yale University Press, 1994.

Howarth, Patrick. *Undercover: The Men and Women of the Special Operations Executive.* London: Routledge and Kegan Paul, 1980.

Hunt, Linda. *Secret Agenda: The United States Government, Nazi Scientists, and Project Paperclip, 1945 to 1990.* New York: St. Martin's, 1991.

Hyde, H. M. *The Atom Bomb Spies.* New York: Ballantine Books, 1981.

Irving, David. *The German Atomic Bomb.* New York: Simon and Schuster, 1967.

——. *Hitler's War.* New York: Viking, 1977.

——. *The Mare's Nest.* Boston: Little, Brown, 1964.

Jacobsen, Annie. *Operation Paperclip: The Secret Intelligence Program That Brought Nazi Scientists to America.* Boston: Little, Brown, 2014.

Jeffreys-Jones, Rhodri. *American Espionage: From Secret Service to CIA.* New York: Free Press, 1977.

Jette, Eleanor. *Inside Box 1663.* Los Alamos, NM: Los Alamos Historical Society, 1977.

Jones, Vincent C. *Manhattan: The Army and the Atomic Bomb*. Washington, DC: Center for Military History, U.S. Army, U.S. Government Printing Office, 1985.

Josephson, Paul. *Physics and Politics in Revolutionary Russia*. Berkeley: University of California Press, 1991.

——. *Red Atom: Russia's Nuclear Power Program from Stalin to Today*. New York: W. H. Freeman, 2000.

Jungk, Robert. *Brighter Than a Thousand Suns: A Personal History of the Atomic Scientists*. New York: Harcourt, Brace, 1958.

Kahn, David. *Hitler's Spies: German Military Intelligence in World War II*. New York: Macmillan, 1978.

Kaiser, David. "The Atomic Secret in Red Hands? American Suspicions of Theoretical Physicists during the Early Cold War." *Representations* 90, no. 1 (Spring 2005): 28–60.

Kaufman, Louis, Barbara Fitzgerald, and Tom Sewell. *Moe Berg: Athlete, Scholar, Spy*. Boston: Little, Brown, 1975.

Keegan, John. *Intelligence in War: Knowledge of the Enemy from Napoleon to Al-Qaeda*. New York: Alfred A. Knopf, 2003.

——. *The Second World War*. New York: Viking, 1989.

Kevles, Daniel J. *The Physicists: The History of a Scientific Community in America*. New York: Knopf, 1978.

Kissinger, Henry. *Nuclear Weapons and Foreign Policy*. New York: Doubleday, 1957.

Knight, Amy. *How the Cold War Began: The Gouzenko Affair and the Hunt for Soviet Spies*. Toronto: McKlellan Stewart, 2005.

Kojevnikov, A. B. *Stalin's Great Science: The Times and Adventures of Soviet Physicists*. London: Imperial College Press, 2004.

Kramish, Arnold. *Atomic Energy in the Soviet Union*. Stanford, CA: Stanford University Press, 1959.

——. *The Griffin*. Boston: Houghton Mifflin, 1986.

——. *The Nuclear Motive: In the Beginning*. Washington, DC: Wilson Center, Smithsonian Institution, 1982.

Kruglov, Arkadii. *The History of the Soviet Atomic Industry*. Translated by Andrei Lokhov. New York: Taylor and Francis, 2002.

Kuhn, Thomas S. "Comment on the Principle of Acceleration." *Comparative Studies in History and Society* 11, no. 4 (1969): 426–30.

——. *The Essential Tension: Selected Studies of Scientific Tradition and Change*. Chicago: University of Chicago Press, 1977.

Kuhns, Woodrow. *Assessing the Soviet Threat: The Early Cold War Years*. Washington, DC: Center for the Study of Intelligence, Central Intelligence Agency, 1997.

Kunetka, James W. *City of Fire: Los Alamos and the Birth of the Atomic Age, 1943–1945*. Englewood Cliffs, NJ: Prentice-Hall, 1978.

Lang, Daniel. *Early Tales of the Atomic Age*. New York: Doubleday, 1948.

Laqueur, Walter. *A World of Secrets: The Uses and Limits of Intelligence*. New York: Basic Books, 1985.

Lasby, Clarence G. *Project Paperclip: German Scientists and the Cold War*. New York: Atheneum, 1971.

Lieberman, Joseph. *The Scorpion and the Tarantula: The Struggle to Control Atomic Weapons, 1945–1949*. Boston: Houghton Mifflin, 1970.

Maddrell, Paul. *Spying on Science: Western Intelligence in Divided Germany, 1945–1961*. New York: Oxford University Press, 2006.

Mahoney, Leo James. "A History of the War Department Scientific Intelligence Mission (ALSOS), 1943–1945." PhD diss., Kent State University, 1981. Microfilm.

Malloy, Sean. *Atomic Tragedy: Henry L. Stimson and the Decision to Use the Bomb against Japan*. Ithaca, NY: Cornell University Press, 2008.

McClellan, James E. III, and Harold Dorn. *Science and Technology in World History: An Introduction*. 2nd ed. Baltimore: Johns Hopkins University Press, 2006.

Moore, Ruth. *Niels Bohr*. New York: Knopf, 1966.

Naimark, Norman. *The Russians in Germany: A History of the Soviet Zone of Occupation, 1945–1949*. Cambridge, MA: Belknap Press of Harvard University Press, 1995.

Newman, James, and Byron Miller. *The Control of Atomic Energy: A Study of Its Social, Economic, and Political Implications*. New York: McGraw-Hill, 1948.

Nogee, Joseph. *Soviet Policy towards International Control of Atomic Energy*. South Bend, IN: University of Notre Dame Press, 1961.

Norris, Robert S. *Racing for the Bomb: General Leslie R. Groves, the Manhattan Project's Indispensable Man*. South Royalton, VT: Steerforth, 2002.

Offner, Arnold. *Another Such Victory: President Truman and the Cold War, 1945–53*. Palo Alto, CA: Stanford University Press, 2003.

Pais, Abraham. *J. Robert Oppenheimer: A Life*. New York: Oxford University Press, 2006.

Parrott, Bruce. *Politics and Technology in the Soviet Union*. Cambridge, MA: MIT Press, 1983.

Parry, Albert. *Peter Kapitza on Life and Science*. New York: Macmillan, 1968.

——. *The Russian Scientist*. New York: Macmillan, 1973.

Paul, Septimus. *Nuclear Rivals: Anglo-American Atomic Relations, 1941–52*. Columbus: Ohio State University Press, 2000.

Pollock, Ethan. *Stalin and the Soviet Science Wars*. Princeton, NJ: Princeton University Press, 2006.

Powaski, Ronald. *March to Armageddon: The United States and the Nuclear Arms Race, 1939 to the Present*. New York: Oxford University Press, 1987.

Powers, Thomas. *Heisenberg's War: The Secret History of the German Bomb*. New York: Knopf, 1993.

Prados, John. *The Soviet Estimate: U.S. Intelligence Analysis and Russian Military Strength*. New York: Dial, 1982.

Price, Don K. *Government and Science: Their Dynamic Relation in American Democracy*. New York: New York University Press, 1954.

Ranelagh, John. *The Agency: The Rise and Decline of the CIA*. London: Weidenfeld and Nicolson, 1986.

Rhodes, Richard. *Dark Sun: The Making of the Hydrogen Bomb*. New York: Simon and Schuster, 1995.

——. *The Making of the Atomic Bomb*. New York: Simon and Schuster, 1986.

Richelson, Jeffrey T. *American Espionage and the Soviet Target*. New York: William Morrow, 1987.

——. *Spying on the Bomb: American Nuclear Intelligence from Nazi Germany to Iran and North Korea*. New York: W. W. Norton, 2006.

——. *The Wizards of Langley: Inside the CIA's Directorate of Science and Technology*. Boulder, CO: Westview, 2002.

Roberts, Sam. *The Brother: The Untold Story of Atomic Spy David Greenglass and How He Sent His Sister, Ethel Rosenberg, to the Electric Chair*. New York: Random House, 2001.

Rose, Paul Lawrence. *Heisenberg and the Nazi Atomic Bomb Project: A Study in German Culture*. Berkeley: University of California Press, 1998.

Rosenberg, David Alan. "The Origins of Overkill: Nuclear Weapons and American Strategy, 1945–1960." *International Security* 7, no. 4 (1983): 3–71.

Rotter, Andrew. *Hiroshima: The World's Bomb*. New York: Oxford University Press, 2008.

Schroeer, Dietrich. *Science, Technology and the Nuclear Arms Race*. Boston: John Wiley, 1984.

Segré, Emilio. *Enrico Fermi, Physicist*. Chicago: University of Chicago Press, 1970.

Sherwin, Martin. *A World Destroyed: The Atomic Bomb and the Grand Alliance*. New York: Knopf, 1975.

Sibley, Katherine. *Red Spies in America: Stolen Secrets and the Dawn of the Cold War*. Lawrence: University Press of Kansas, 2004.

Sime, Ruth Lewin. *Lise Meitner: A Life in Physics*. Berkeley: University of California Press, 1996.

Smith, Alice Kimball. *A Peril and a Hope: The Scientists' Movement in America, 1945–47*. Chicago: University of Chicago Press, 1965.

Smith, R. Harris. *OSS: The Secret History of America's First Central Intelligence Agency*. New York: Dell, 1973.

Snow, C. P. *Science and Government*. Cambridge, MA: Harvard University Press, 1960.

Soderqvist, Thomas, ed. *The Historiography of Contemporary Science and Technology*. Amsterdam: Harwood Academic, 1997.

Soyfer, Valery. *Lysenko and the Tragedy of Soviet Science*. New Brunswick, NJ: Rutgers University Press, 1994.

Strickland, Donald. *Scientists in Politics: The Atomic Scientists Movement, 1945–1946*. Lafayette, IN: Purdue University Studies, 1968.

Sylves, Richard T. *The Nuclear Oracles: A Political History of the General Advisory Committee of the Atomic Energy Commission, 1947–1977*. Ames: Iowa State University Press, 1987.

Thiesmeyer, Lincoln, and John Burchard. *Combat Scientists*. Boston: Little, Brown, 1947.

Thomson, Sir George. "Anglo-U.S. Cooperation in Atomic Energy," *Bulletin of the Atomic Scientists* 9, no. 2 (1953): 46–48

Thorpe, Charles. *Oppenheimer: The Tragic Intellect*. Chicago: University of Chicago Press, 2006.

Troy, Thomas. *Donovan and the CIA: A History of the Establishment of the Central Intelligence Agency*. Washington, DC: Center for the Study of Intelligence, Central Intelligence Agency, 1981.

Volti, Rudi. *Society and Technological Change*. New York: Worth, 2001.

Walker, J. Samuel. *Prompt and Utter Destruction: Truman and the Use of Atomic Bombs against Japan*. Chapel Hill: University of North Carolina Press, 2004.

Walker, Mark. *Nazi Science: Myth, Truth, and the German Atomic Bomb*. Cambridge, MA: Perseus, 1995.

Wang, Jessica. *American Science in an Age of Anxiety: Scientists, Anticommunism, and the Cold War*. Chapel Hill: University of North Carolina Press, 1999.

Warner, Michael, ed. *The CIA under Harry Truman*. Washington, DC: Center for the Study of Intelligence, Central Intelligence Agency, 1994.

Weart, Spencer. *Scientists in Power*. Cambridge, MA: Harvard University Press, 1979.

Weart, Spencer, and Gertrud Weiss Szilard, eds. *Leo Szilard: His Version of the Facts*. Cambridge, MA: MIT Press, 1978.

Weinstein, Allen, and Alexander Vassiliev. *The Haunted Wood: Soviet Espionage in America; The Stalin Era*. New York: Random House, 1999.

Wells, Herbert George. *The World Set Free: A Story of Mankind*. New York: E. P. Dutton, 1914.

Williams, Robert Chadwell. *Klaus Fuchs, Atom Spy*. Cambridge, MA: Harvard University Press, 1987.

Williams, Susan. *Spies in the Congo: America's Atomic Mission in World War II*. New York: Public Affairs, 2016.

Wittner, Lawrence. *One World or None: A History of the World Nuclear Disarmament Movement through 1953*. Palo Alto, CA: Stanford University Press, 1993.

Wolfe, Audra. *Competing with the Soviets: Science, Technology, and the State in Cold War America*. Baltimore: Johns Hopkins University Press, 2012.

Zaloga, Steven. *Target America: The Soviet Union and the Strategic Arms Race, 1945–1964*. Novato, CA: Presidio, 1993.

Ziegler, Charles, and David Jacobson. *Spying without Spies: Origins of America's Secret Nuclear Surveillance System*. Westport, CT: Praeger, 1995.

Zimmerman, David. *Top Secret Exchange: The Tizard Mission and the Scientific War*. Montreal: McGill–Queen's University Press, 1996.

Zubok, Vladislav, and Constantine Pleshakov. *Inside the Kremlin's Cold War: From Stalin to Khrushchev*. Cambridge, MA: Harvard University Press, 1996.

INDEX